INTRODUCTION TO
MECHANICAL RELIABILITY

INTRODUCTION TO MECHANICAL RELIABILITY: A Designer's Approach

Oleg Vinogradov

Department of Mechanical Engineering
University of Calgary
Alberta, Canada

●HEMISPHERE PUBLISHING CORPORATION

A member of the Taylor & Francis Group

New York   Washington   Philadelphia   London

INTRODUCTION TO MECHANICAL RELIABILITY: A Designer's Approach

1 2 3 4 5 6 7 8 9 0   B R B R   9 8 7 6 5 4 3 2 1

This book was set in Times Roman by Graphic Composition, Inc.
The editors were Carol Taylor Edwards and Carolyn V. Ormes.
Cover design by Sharon M. DePass. Printing and binding by Braun-Brumfield.

A CIP catalog record for this book is available from the British Library.

Library of Congress Cataloging-in-Publication Data

Vinogradov, Oleg (Oleg G.)
    Introduction to mechanical reliability : a designer's approach /
Oleg Vinogradov.
        p.   cm.
    Includes bibliographical references and index.

    1. Reliability (Engineering) I. Title.
    TA169.V56   1991
    620′.00452—dc20   90-24181
                                                            CIP

ISBN 0-89116-920-2

To my wife Aleksandra and son Mark

CONTENTS

Preface   ix

1 Mechanical Reliability   1

1-1   Introduction   1
1-2   Uncertainties in Mechanical Engineering   6
1-3   Reliability Engineering   11
1-4   Reliability and Probability   14
      Problems   21

2 Integrating Reliability into Design   23

2-1   Introduction   23
2-2   System Reliability   24
2-3   Conceptual Design Stage   26
2-4   Preliminary Design Stage   43
2-5   Detailed Design Stage   56
      Problems   65

3 Testing for Reliability   73

3-1 Introduction   73
3-2 Types of Tests   73
3-3 Design of Experiments   76
3-4 Alternative Testing Techniques   86
      Problems   93

4 Operational Reliability   95

4-1 Introduction   95
4-2 Failure Causes and Trends   95
4-3 Some Mathematical Models of Failure Rates   99
4-4 Life Data Analysis Versus Design   114
      Problems   117

5 Statistical Simulation   119

      Problems   128

Appendixes

A Bibliography   131
B Statistical Tables   133
C Weibull Distribution Probability Paper   139

Index   141

PREFACE

In spite of the number of excellent books already available on various aspects of reliability engineering, I feel that an introductory text is needed in which reliability is treated as part of the design process, bridging an interface between reliability and design so that communication between various people in a team is more efficient. Such a feeling comes from my experience in industry and in teaching undergraduate mechanical engineering. My intention is to make the book concise, clear, and illustrative, so that the reader develops an overall understanding of the problems, concepts, and methods of reliability engineering without going into particulars. I am, however, aware of the danger of being superficial when trying to combine conciseness and clearness. It is left to the judgment of the reader to decide whether the balance is achieved.

Although the book is intended to be self-contained, a familiarity with the basics of probability and statistics is assumed. A number of examples are given to illustrate the theory and to develop in the reader a "feeling" for the subject. However, the reader is warned that these are only examples and in each real-life case, personal judgment should be exercised. It should also be noted that the book is not an extension of probability and statistics into reliability engineering, rather it is reliability based on concepts of probability and statistics.

The selection of topics is not an indication of their importance in reliability engineering but rather the reflection of my view on what might be most valuable for a beginner. In Chapter 1, types of uncertainties in mechanical engineering are considered, then the notion of reliability as no-failure performance is introduced, and finally, reliability measures are discussed. In Chapter 2 the integration of reliability into the design at various stages is considered. It is shown that conceptual, preliminary, and design development stages all have their own objectives in general and, in terms of reliability requirements, in particular. Chapter 3 discusses the effect of testing (component or prototype) on product reliability. The number of specimens, the conditions of testing, and the interpretation of results are considered. In Chapter 4 the assessment of reliability of the product in service is discussed. The emphasis is on data gathering, analysis, and interpretation of results. The inclusion of Chapter 5 is justified by the fact that mechanical engineers play an important role in the statistical simulation of system performance. In order to cooperate efficiently with reliability experts a mechanical engineer should understand the objectives, the input information needed for simulation, and the technique of simulation in general terms.

This book is intended for undergraduate students of mechanical engineering, for whom it will serve as a natural extension into unconventional aspects of machine design, and for practicing engineers not specializing in reliability engineering.

Oleg Vinogradov

CHAPTER ONE

MECHANICAL RELIABILITY

1-1 INTRODUCTION

The development of an engineering product is a process with several distinct stages. This process is analogous to a flowchart with the logical "ifs" at the end of each stage, when it is decided whether the process should continue, be returned to a previous stage, or be stopped. This book examines the role of reliability in the process of product development and the decision making involved.

Any product is developed to meet specific functional objectives, for example, to convert energy from one form to another, to fly, to detect flaws. These functional objectives indicate the need for a product. The associated task is to specify all the conditions that the product must satisfy in order to be competitive or operational. These conditions are design constraints, such as manufacturing cost, safety requirements, and weight and size limitations. The formulation of functional objectives and design constraints is done before the product development process starts. Continuing the flowchart analogy, this initial step constitutes an input into the process.

The design process begins by analyzing possible ways of achieving the functional objectives of the product. The various physical arrangements resulting in, theoretically, the same functional outcome are called the alternatives. The objective, at the start of the design process, is to develop various alternatives and then to compare them in order to choose one or several that, based on experience or estimations, promise to satisfy the imposed design constraints. The emphasis at this stage of product development is on a concept rather than on meeting constraint requirements. Therefore it is often called the *conceptual stage* of product design. Note that the notion of concept in this context includes both the physical principles that form the basis for achieving the functional objectives and an outline of the product configuration.

Example 1-1 There is a need for a diesel locomotive able to pull a train of a given load within a specified range of speeds (the functional requirement of the locomotive). First, the cost to develop such a locomotive has an upper limit (otherwise the railroad would not be able to compete with other means of transportation). This is the cost constraint. The second constraint is caused by the maximum allowable contact pressure between each wheel and the rail. A given wheel and rail geometry places limits on the weight/wheel ratio. This is the weight constraint. Then the power requirements must be satisfied. The power requirements follow from the functional requirements. However, the needed power can be supplied by a single engine or by multiple-engine arrangements, depending on the availability of engines and on safety considerations. Availability and safety are two other constraints, and so on. Thus the functional objective of pulling a specific train at a specified speed can be achieved by many possible locomotive configurations.

One or more alternatives may be chosen for further development, and then they enter the *preliminary design* stage. While during the conceptual stage the emphasis was mostly on satisfying the functional requirements, during the preliminary design stage, it shifts to meeting the imposed constraints. Essentially, the objective of this stage of design is to show that the product with the specified requirements and constraints can indeed be made. This is done by performing theoretical analyses of elements critical to the product functioning and by working out product layouts clarifying the overall dimensions and the product configuration. At this stage a more specific input of information on product functioning is required, namely, details of loads, speeds, temperatures, and environment. On the basis of this more in-depth design a decision on further product development is made.

Example 1-2 The product is a transmission gearbox. Its kinematic diagram defining the velocity ratio was determined during the conceptual design stage. During the preliminary design stage, the type and dimensions of gears, shafts, bearings, and clutches are determined, based on the power and speed requirements. Also, the lubrication requirements and ways of meeting them are worked out. Based on this information, a general view of the gearbox, defining overall dimensions, is made. At this point an approximate estimation of weight is possible and cost assessment becomes more specific.

Usually, only one alternative enters the *detailed design* stage, at which a complete set of working drawings is produced and specifications are outlined. The objective of this stage is to present the design in such a form as to show how to make it. Dimensions of all components are finalized at this time. All accompanying drawings requirements, such as the accuracy of manufacturing, the materials used and their properties, the types of lubricants, and the components supplied by vendors are called specifications. In order to formulate these specifications, a more in-depth theoretical analysis of stresses, heat transfer, lubrication, vibrations, and so on is required. If some important elements of the design cannot be analyzed theoretically, then a test or a series of tests simulating their performance is needed. In some cases, theoretical analysis, computer simulation, and experimental investigation are all done simultaneously, or sequentially, if a particular design element is critical to the product functioning.

Example 1-3 The blades of a steam turbine are subjected to centrifugal force (as well as aerodynamic forces) at elevated temperatures. For a low-pressure part of the turbine, the blades can be so long that the turbine safety due to blade elongation and high stresses becomes critical. The blade is a three-dimensional body having a very complicated shape. Although computer-based theoretical methods allow analysis to be made, the properties of the material under elevated temperatures and various stress levels may not be well known. Then, in addition to theoretical analysis, a prototype of the blade is made and tested in a temperature and stress environment similar to that under working conditions.

At the end of the detailed design stage the product is ready to enter the *manufacturing stage*. The objective of this stage is to manufacture the product in conformance with the specifications outlined in the design. Correspondingly, decisions should be made with respect to equipment and tools to be used, methods of production, and quality control system. The assembling and, sometimes, the idle running are also part of the manufacturing stage.

When the product is assembled, it is ready for the *service stage*. In short, all previous stages can be called design and development. Product life starts when it enters the service stage. For most products, service (use) does not start immediately after assembly because of the length of time required for shipping and storage. Nevertheless, when in service, the product will be expected to perform as anticipated, i.e., according to the specified functional requirements. The objective for the service stage is to maximize product life while keeping the products' functional characteristics within allowable limits.

The preceding gives a simplified scenario of product development and service. In reality, however, it is not a sequential process but an iterative one with continual feedback due to growth of knowledge and information. The number of stages can also differ, depending on previous experience and product complexity. What is important, however, is that the indicated sequence of stages breaks down the product development and service cycle into qualitatively different intervals, each having its own methods of meeting its objectives.

Product development is a decision-making process, and reliability constitutes one of the aspects that decisions are based on. Reliability as a concept needs an explanation. Usually, it is associated with the level of confidence in the product. However, it is not a subjective but rather an objective attribute of the product. Reliability is represented by a number indicating, for example, the probability that the product will satisfy functional requirements during a specified period of time without failure. Failure means any event that prevents the product from fulfilling its functional objectives. Accordingly, not only catastrophic breakages but also deterioration of properties below an acceptable level, i.e., vibrations, loss of efficiency, or leakage may constitute failure. The product design process goes hand in hand with the reliability growth process. Generally, neither manufacturing nor service stages can increase the initial level of reliability. *The reliability of the product is built into its drawings and specifications.* Manufacturing does not change specifications; it may, however, not conform to them, thus reducing the reliability level. In this respect it is important to understand how reliability is built into the product during the various design stages.

At the conceptual design stage, reliability is not usually a main concern, since no product has been conceived for the sake of reliability alone. Nevertheless, this is the time when the foundation of reliability is laid because basically the product configuration, determined at this stage, plays an important role in reliability. Overall reliability can be increased by choosing a better arrangement of components irrespective of the reliability of the components themselves. In Example 1-1, it would be decided at this stage how many wheel pairs, or engines, or cooling devices to have per locomotive. The decision concerning product configuration should be based on clearly stated reliability considerations as well as some other constraints. At the same time, any decision concerning reliability of components would be based on the experience of the design engineer and would be expressed by such words as "realistic," "reasonable," or "sound." It is important to realize that, while the reliability of the product scheme can be quantified using a relative scale, the reliability of components at this stage is assessed only qualitatively. Still, some elements have more designer's confidence than others. This confidence forms the ground for formulating reliability concerns and for outlining reliability analysis to be performed during the following stages of design.

At the preliminary design stage the substance of reliability assessment changes. Two levels of such an assessment are in use: one is based on a probabilistic approach and the other is a traditional estimation based on the safety factor. Such a coexistence of various approaches is a natural phenomenon at this stage when two or more alternatives are competing. Only elements critical to the survival of each alternative should be thoroughly analyzed at this stage. Nevertheless, estimations of stress, force, speed, flow rate, and temperature all increase the level of confidence in the product. The reliability, although not yet quantified in its strictest sense, grows. For example, the dimensions of gears, bearings, and shafts in Example 1-2 are found using information (or guessing) on the maximum loads, approximate properties of materials, and surface hardnesses. A more in-depth analysis is postponed until an alternative enters the detailed design stage.

At the detailed design stage, when there are no, or almost no, doubts that the concept is correct and that it can be materialized, the goal is to foresee all possible service situations and to develop the design accordingly. Thus the emphasis is on product reliability. The development of working drawings proceeds in parallel with the analysis of product performance based on finalized dimensions, materials, and configurations, and with experimental investigations of critical components or subsystems. As a result, the reliability of the product is finalized. However, at this stage, the real magnitude of reliability in most cases remains unknown. What is known is its estimation, which is as accurate as the designer's prediction of the environmental conditions under which the product will operate and the ability of the product to withstand those conditions. The important point is that reliability is an objective attribute of the product, which is only revealed in service. The discrepancy between the designer's prediction of reliability and the real life behavior of the product is a reflection of the discrepancy between the state of knowledge and reality. So the product leaves the design stage with some built-in level of reliability.

During the manufacturing stage the reliability, at best, is not decreased if the design specifications are complete and are met. However, the violation of specified requirements is possible even under the best quality control system because the design specifications may be vague and open to interpretation by the industrial engineer, especially if the communication link between the industrial engineer and the design engineer is weak or nonexistent. Also, manufacturing processes may vary considerably, decreasing the level of reliability built into the product during the design stage. Note that the reliability reduction factor due to malfunction in manufacturing is controlled by the quality control personnel and not by the design engineer.

As far as product reliability is concerned, two more stages are sometimes important and may contribute to the reduction of reliability: storage and transportation. Aging of lubricants, oxidation of contact surfaces, and corrosion of materials during storage and severe loads or environmental conditions during transportation may result in the possibility of product malfunction while in service. Again, decisions concerning storage and transportation conditions that affect product reliability are made by different groups of people. If a communication link does not exist between these decision makers and the design engineer, there may be dire consequences.

In service, a distinction should be made, from the reliability point of view, between products with and without maintenance. For a product without maintenance the reliability is predetermined during the design, manufacturing, storage, and transportation stages. The only remaining task is to collect data on product performance to improve prediction techniques and to develop more realistic expectations on product life. Sometimes products without maintenance are called "one-shot" systems, likening them to a shell or missile.

For a product with maintenance the situation is much more complicated. First, a distinction should be made between preventive maintenance and corrective maintenance. The former, designed to extend product life, represents an active control of product reliability. The latter is forced by an unexpected failure. The downtimes of the product due to maintenance or repair are also different: while preventive maintenance is done during times planned in advance, corrective maintenance is unpredicted, causing forced downtime. Any downtime constitutes a loss in product performance and, in many situations, results in the need for more redundant products. For example, to maintain a passenger flow requirement, the identical or nearly identical number of planes should be operational; to maintain the required level of energy production, a specific number of energy units should be operational, and so more redundant planes or energy units are needed if downtimes occur more often. Thus, even if the level of product reliability is restored after maintenance, the duration of downtimes is a reflection on the ability of the product to fulfill its functional requirements within a specified time interval. The duration of the forced downtime depends on the type and location of the failure and the availability of spare parts, machine tools, and qualified personnel. It should be noted that all these parameters are independent of product reliability. It follows that the duration of the forced downtime is a characteristic independent of reliability. The number of downtimes and their duration affect the efficiency of the product and the maintenance costs.

It is clear, then, that the overall forced downtime during the product service becomes an important factor characterizing any product. This characteristic is called maintainability. The smaller the total forced downtime for a product, the better its maintainability characteristic. The relationship between product reliability and maintainability should be clearly understood. The reliability of the product is independent of its maintainability if, after the breakdown, the level of reliability is restored. In practice, however, this is not the case, since the number of forced shutdowns affects the rate of degradation of product reliability. However, the maintainability is closely related to the reliability of the product, since more failures cause more downtimes. So, although the duration of each downtime is a characteristic independent of reliability, the number of downtimes during the product life follows from product reliability. In order to achieve a highly effective product, both high reliability and maintainability are required. *The main objectives of reliability engineering are both to predict and to maximize the effectiveness of product performance.*

As a subject, reliability engineering is independent of the other aspects of product development and utilization. However, as a constraint imposed on a product, it is closely related to other constraints and considerations. For example, a reliability constraint is inseparable from cost considerations. High reliability results from more in-depth analysis, more tests, better manufacturing equipment and quality control, a better system of preventive maintenance, and a larger stock of spare parts; in short, it means high costs. On the other hand, low reliability means more failures, extended downtimes, reduced productivity, and more reclamations: in short, reduced revenues. The difference between high and low reliability may be the difference between a profitable and a nonprofitable enterprise. For each product and market situation there is a balance between the two, resulting in a maximum profit (see Fig. 1-1). However, there are situations when nonmonetary considerations may prevail, when safety becomes a dominant criterion.

It follows from the above that the reliability requirements cannot be considered independently from other aspects of product development and implementation. Rather, the reliability requirements should be formulated as a result of studying other aspects of product life such as safety, costs, market, and productivity. Formulation of product reliability requirements is a difficult task if reliability is not a dominant criterion. In some industries, like defense, aerospace, or electronics, the reliability requirements are part of the product specifications; in consumer-oriented industries, however, they may not exist at all. In any case, it is important to realize that sometimes all that is required to prevent substantial losses is a basic knowledge of reliability engineering.

1-2 UNCERTAINTIES IN MECHANICAL ENGINEERING

Usually, the functional objectives and imposed constraints of the product are expressed in numbers; for example, the projected efficiency is 85 percent, the maximum speed is 160 km/h, the output power is 200 hp, and the lifespan is 5000 hours. These are called the nominal parameters of the product. In reality, these parameters can rarely be achieved exactly, since there is always some possibility for their scatter. Scatter of product functional and constraints parameters is a fundamental law of nature. Since scatter cannot be eliminated, the objective of engineering science is to study what causes the scatter, the range of scatter, the methods of control, and how to take scatter into account at the design, manufacturing, and service stages.

Scatter of product parameters is caused by the inability of existing technology to reproduce the same item absolutely identically and by the lack of knowledge concerning various natural and man-made phenomena. For example, for the same grade of steel, properties such as yield stress, endurance strength, and creep will inevitably

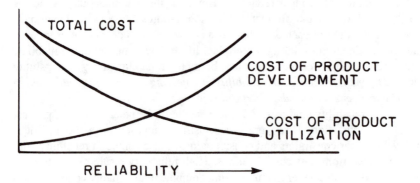

Figure 1-1 Qualitative cost-reliability relationship.

vary due to small variations in chemical composition, temperature in the furnace, and nonhomogeneous temperature field during solidification. No two shafts can be made identical due to tool wear, errors in measurement, machine tool vibrations, resistance of the shaft material to cutting, and so on. The efficiency of a hydraulic turbine varies due to the sensitivity of turbine performance to the blade geometry, spacing of blades, quality of blade surfaces, and nonhomogeneity of velocities at the entrance of the turbine. The scatter of the load on a washing machine is, besides other factors, only a reflection of scatter of consumer attitudes. Scatter of product parameters means that for any particular product it is impossible to predict its properties with certainty. This has always been recognized in engineering. However, only relatively recently, in the second half of this century, has the gray area of uncertainty become the object of intensive study.

All uncertainties in reliability can be divided into two categories: physical, associated with the state of nature, and cognitive, associated with the state of knowledge. More emphasis is given below to explain the meaning of physical uncertainty, while cognitive uncertainty is discussed only briefly and is left to be clarified in later chapters.

Physical uncertainties can be subdivided into those that are internal, associated with the properties of the product itself, and those that are external, associated with the environment in which the product operates.

Internal physical uncertainties are caused by the variability of technological and operational processes. The design engineer takes this variability into account when assigning manufacturing tolerances, limits of material properties, and boundaries for productivity and efficiency variations. By making these types of decisions, the design engineer is establishing the ranges within which the design parameters might vary. This step is one of the most important in the design process, since the range of variability of product properties depends upon it. However, the connection between the variability of the parameters of product components and the variability of product properties is not straightforward. The study of this relationship is one of the objectives of probabilistic design. It should be noted that although the range of parameter variability is a subjective attribute, since it is formulated as a requirement in the specifications, the variability itself is an objective attribute, since it reflects the laws of nature. The following examples are included to demonstrate the existence of internal physical uncertainties and their role.

Example 1-4 A press fit (Fig. 1-2) is intended to prevent slippage, while torque T is transmitted from the shaft to the gear. A mathematical expression for the torque transmitted by the press fit is

$$T = \frac{1}{2} \pi f p L D^2 \tag{1-1}$$

where D is the diameter of the shaft, L is the thickness of the gear disk, p is the contact pressure, and f is the coefficient of friction. The contact pressure is a function of disk and shaft interference Δ and is estimated by

$$p = \frac{E}{4R} \left[1 - \frac{1}{4}\left(\frac{D}{b}\right)^2 \right] \Delta \tag{1-2}$$

where E is the modulus of elasticity and b is the disk diameter.

In general, all dimensions vary within the limits specified by the tolerances; the modulus of elasticity also varies. As a result, the holding ability of the press fit varies within certain limits. Since

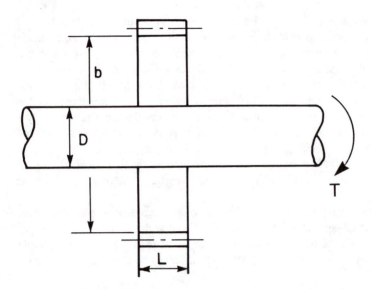

Figure 1-2 A shaft–gear press fit coupling.

tolerances are assigned to all parameters, they all appear to play an equally important role in product performance. However, in engineering analysis, it is important to identify those parameters with variations that most affect performance. Usually, these are the parameters with relatively wide scatter. In the case under consideration, the modulus of elasticity and geometrical parameters D, b, and L differ little with respect to the nominal values, whereas the effect of the coefficient of friction and the interference of tolerances is much more pronounced.

Friction is strongly dependent on the geometry of mating parts, namely, on the real contact area at the interface. Because of imperfections in manufacturing controlled by the tolerances on the roundness of a cross section, the straightness of the shaft, and surface roughness (wavelike variations with superimposed fine irregularities), the real area of contact might vary significantly in relation to the nominal contact area. This variation is equivalent to the variation of the coefficient of friction. The coefficient of friction also depends on the surface condition—clean, dry, lubricated—and on the strength of the pressure, since the higher the pressure, the larger the real area of mating parts that come in contact. It follows that the value of the coefficient of friction depends on many random factors and remains unknown.

Another factor strongly affecting the holding ability of the press fit is the interference of tolerances, illustrated in Fig. 1-3. The variation of the shaft and disk diameters within specified tolerances results in the variation of the interference between minimum and maximum. The main conclusion is that the torque-carrying ability of the press fit varies within unknown limits.

Example 1-4 also demonstrates how the interaction of components with uncertain properties leads to uncertain properties in the assembly. The following example stresses, once more, the influence of assembly on the scatter of product parameters.

Example 1-5 Two dynamically balanced shafts are to be coupled together by bolt joints as shown in Fig. 1-4a. Note that in any ideally balanced rotor the centers of gravity in each cross section coincide with the axis of rotation. The question is whether the assembly remains balanced and, if it is not, how to predict the unbalanced forces. The answer to the first question is negative due to the allowed imperfections in manufacturing. First of all, the surfaces of the flanges are not ideally perpendicular to the corresponding axes of rotation (strictly speaking, it means that the shaft itself cannot be ideally balanced, but this effect is assumed to be negligible for the sake of simplicity). The second factor is that

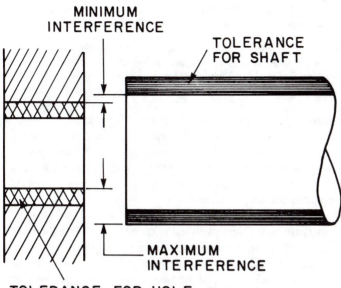

Figure 1-3 Illustration of tolerances and interferences in a press fit.

the flange surface is not exactly a plane. As a result of both factors, the angular misalignment of assembled shafts is possible.

Another type of misalignment (parallel) is caused by the bolt joints. The holes in the flanges are usually made larger than the bolt diameters in order to compensate for the tolerances on the radial and pitch positions of the holes, unless they are drilled together. As a result, two shafts with the bolts untightened remain loose with respect to their transverse displacement. The final possible geometry

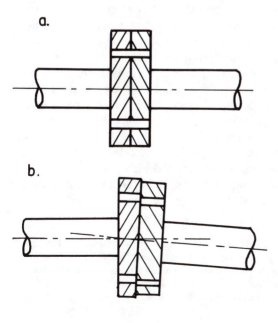

Figure 1-4 Effect of coupling on a misalignment of the assembly: (*a*) ideal coupling and (*b*) misalignment caused by nonperpendicularity of one of the flanges and eccentricity of the bolt joint.

after assembly is shown in Fig. 1-4*b* in an exaggerated form. It is clear that the centers of gravity do not coincide with the center of rotation. This is why the static or dynamic balancing of assembled rotors is required.

Since unbalanced forces cause vibrations in rotating systems, the level of vibrations, as well as the mode of vibrations due to random locations of unbalanced forces, can vary within limits and be larger than the acceptable level. Prediction of the unbalanced forces is a typical problem of probabilistic design.

In general, by assembling the components into the product, new uncertainties arise, causing unbalanced forces, friction forces, thermal forces, and unequal distribution of forces between bolts, keys, or splines. In addition, an operating product creates its own internal environment, temperature, humidity, and chemistry, which is the result of internal processes and interaction with the surrounding environment. These are internal uncertainties due to interaction.

Examples 1-4 and 1-5 also demonstrate that in each particular situation it is important to identify the most critical parameters either to keep them under tighter control or to study their potential effect. Identification of the critical parameters is usually done by the design engineer, while the influence on system performance is monitored by the reliability engineer. Cooperation between the two engineers becomes very important.

External uncertainties are associated with variability in either operational or environmental conditions, or both. Since the limits of operational conditions can be specified, it follows that the corresponding external uncertainties, in most cases, have an upper limit and can be controlled, i.e., the maximum load a truck can carry, the lowest ambient temperature a car radiator can withstand, or the maximum ice thickness that a vessel can negotiate. Uncertainties associated with environmental conditions also have an upper limit, but contrary to operational uncertainties, their upper limits remain in many cases unknown and cannot be controlled. Winds, waves, earthquakes, ambient temperatures, and humidity are all environmental uncertainties.

In many situations a combination of operational and environmental uncertainties occurs. For example, if the same type of excavator was used to dig sand in the South and clay mixed with rocks in the North, the difference in the operational and environmental conditions would be significant. One important conclusion follows from this example: there are situations when the design engineer is not fully aware of the conditions in which his design is going to be used and thus his operational and maintenance specifications (manuals) cannot be fully adequate. One of the objectives of life data analysis in reliability engineering is to study this relationship between operational and maintenance requirements and environmental conditions. If the design engineer receives feedback, then the new information on product performance can be used to adjust the operational and maintenance requirements to specific conditions.

Limited or unlimited, controlled or noncontrolled, the operational and environmental conditions result in uncertainties that the design engineer faces while making decisions.

Example 1-6 The lifetime of a standard ball bearing depends, all other factors being equal, on speed and load. Let us assume that the bearing is a part of a car transmission gearbox and that all bearings of this type are identical. The design engineer has to make the decision when to replace this bearing with a new one. The limiting speeds and loads are known to the designer, who has two options: first, to presume extreme operating conditions and, second, to take into account the variability of speeds and loads. The first approach would result in overdesign, since the bearing that is not yet worn out would be replaced. The second approach might extend the lifetime of the bearing. However, if the analysis in the second approach is not done properly, it might result in failure, with unpredictable consequences. Two factors are important with respect to this problem: (1) the existence of adequate statistics for similar gearboxes operating in similar conditions, and (2) the availability of a method for predicting the lifetime of a ball bearing based on its operational statistics.

Cognitive uncertainty is associated with our interpretation of the physical world and includes sampling uncertainty, model uncertainty, and computational uncertainty. Usually, a set of identical products used under similar conditions is called a population, a subset of which is called a sample. A sample drawn from the population is used to make statistical inferences about the entire population. The discrepancy between the true statistics characterizing the population and those based on the limited sample is called the sampling uncertainty.

Any given statistic is approximated by an appropriate probabilistic model. Since such a model is always based on some assumptions that cannot be verified unless more data or more complicated models are analyzed, this results in another uncertainty, called the model uncertainty. Implementing the probabilistic model into design might require some additional assumptions allowing the utilization of the developed computational methods. Again, the limitations imposed by such assumptions often remain vague, resulting in another uncertainty, called the computational uncertainty.

The nature of reliability is associated with physical uncertainty, whereas cognitive uncertainty is due to lack of information. Nevertheless, both contribute to uncertainty in predicting the reliability of a product.

Service brings into play another type of uncertainty affecting the productivity or utilization of a product, downtime uncertainty, associated with the duration of the forced downtime due to failure. The duration of the forced downtime is difficult to predict, since the location of the failure, its accessibility, the availability of spare parts or machine tools, and the availability of skilled and experienced personnel can differ for the same product used by different customers. The duration of the forced downtime can vary, but what is important is that it cannot be planned in advance and it is random in nature. It depends, among other things, on the organization of maintenance services and may differ from one enterprise to another. Thus, each enterprise will generally have its own statistics and statistical and model uncertainties. The important point is that the downtime uncertainties are usually impossible to predict at the design stage, since they occur only in service. The statistics needed to establish better maintenance schedules and procedures are gathered as a result of inadequate maintenance schedules and procedures. The question is how to set up a proper maintenance system while incurring minimum losses. This is the objective of maintainability engineering.

1-3 RELIABILITY ENGINEERING

Uncertainties associated with the product and with the environment in which it operates lead to uncertainty in product performance. Reliability engineering is concerned with the prediction of the period of time during which a product has acceptable functional properties. The formal definition of reliability used in this text is as follows: the reliability of a product is the probability that in given environmental and operational conditions it will perform its function without failure for a specified period of time.

The word "probability" in the above definition indicates that the problem of prediction is considered in a statistical sense. The clause "given environmental and operational conditions" identifies constraints imposed on the utilization of the product whose reliability is estimated, namely, it is assumed that these conditions remain identical for all similar products under consideration. For example, two fleets of similar cars, one operating in a cold and humid climate, the other in a warm and dry climate, cannot be compared from the viewpoint of reliability, since they are subjected to different environmental conditions. Extending this example, the life data will also be incompatible if the two fleets are subjected to different operational conditions, one on relatively flat terrain, the other in a rugged or mountainous area. The notion of similar cars in this example should be further clarified. Cars, or any other product in this respect, can be considered similar in the context of reliability assessment if they are made according to the same design specifications, using the same technological processes, subjected to the same quality control system, and shipped and stored under similar conditions.

A set of similar products operating under similar environmental and operational conditions is called a population. It is clear from the above that for any product the statistics are meaningful only if they describe the same population.

Finally, the notion of failure requires elaboration. A failure is defined here as an event whose occurrence results in either the complete stoppage of an operation or in an unacceptable deviation of operating characteristics. This definition does not include an explanation of how the product came to a state of failure in the first place. Although reliability engineering deals with failures as discrete events in time, the prediction of reliability very much depends on the process of arriving at the state of failure. From this point of view, all failures can be divided into two broad categories: instantaneous and gradual. Instantaneous failure denotes a failure due to an abrupt change in the "load-resistance" relationship, whereas gradual failure denotes failure due to a gradual change in that relationship. Note, the combination of the qualifying word "gradual" with the word "failure" is used in order to reflect the history of arriving at the state of failure; otherwise it is contradictory, since a failure as an event is always instantaneous. Instantaneous failures are due to overloading and may result in fracture, buckling, or seizure, whereas gradual failures are mostly due to fatigue, corrosion, creep, or wear. However, gradual deterioration of properties often results in instantaneous failure. In the following, the load-resistance concept should be understood in a broad sense, with the load encompassing all environmental and operational conditions, and with the resistance implying the ability to withstand them for a specified period of time.

There are various combinations of load-resistance characteristics that are important to distinguish in reliability engineering: (1) both load and resistance are time-independent, (2) one is time-independent, and (3) both are time-dependent.

If both load and resistance are time-independent, then strictly speaking, it is not a problem of reliability engineering, which is concerned with a failure-free operation during a specified period of time. In this case, however, a failure may take place only at the moment the product is put into operation. Still, uncertainties associated with both load and resistance are present, which means that the assessment of failure at the

beginning of operation can be done only in probabilistic terms. This is a true probabilistic design problem. A case in point is a pressure vessel under a static load when
the deterioration of the material of the vessel is prevented. Note that the time independence of load and resistance can be assumed to be valid only for a set period of time,
that is, from one maintenance to another, otherwise it leads to an infinite life.

If the load is time-independent (i.e., environmental and operational conditions do
not change) but the resistance changes with time, then the prediction of time to failure
depends on the rate of decay of product properties. In most cases this rate is difficult
to predict with certainty, since it depends on many factors associated with internal
uncertainties.

Example 1-7 A steam turbine generates the same amount of energy during a long period of time and
operates with constant speed in a similar environment. The load seems to be, in this case, time-
independent. The resistance, however, changes. Consider a journal bearing. Its functional properties
depend on the geometry of the journal and the bearing, the viscosity of the lubricant, the flow rate of
the lubricant, and an unbalanced dynamic load on the bearing. Due to deterioration of the oil, leading
to its decreased viscosity and increase in wear, all parameters mentioned above do not remain constant
but gradually change with time. The rate of this change depends, other things being equal, on the
values of the above-mentioned parameters at the beginning of the observation. Since initial conditions
are randomly scattered within a certain range, so is the rate of deterioration. It should be noted that
deterioration of the bearing will result in an increased level of vibration. If the latter is monitored, then
the failure of the bearing can be prevented.

To give another example, a combination of constant resistance and varying load
is often encountered in practice when the rate of decay of product properties is much
slower than the rate of change of environmental or operational conditions, so that the
former can be assumed to be constant. In this case failure occurs at that point in time
when the load becomes larger than the resistance. If the load-time relationship is
known then, for a given resistance, it is a problem of reliability engineering. Unfor-
tunately, the load-time relationship is seldom known with certainty. In reality, either
the magnitude of the load or the time factor is a random parameter, or both are random.
For example, in the case of an unbalanced shaft the magnitude of the load is random,
while the time factor can be deterministic; in transportation in many situations the
tractive force can be considered to be deterministic, while the regime of operation (the
time factor) is random; in the case of ship-ice interaction the magnitude of the forces
and the time of their action are both random.

In general, both load and resistance are time-dependent. The situation becomes
even more complicated when resistance is affected by the load, as in the case of fatigue
or wear. Usually the resistance is a steadily decreasing function of time. For example,
the resistance of a shaft subjected to cyclic bending gradually decreases due to the
origination and propagation of cracks; the resistance of a ball bearing decreases due to
the increase of the viscosity of the lubricant and deterioration of the initial geometry
in the ball-trace contact zone because of wear, seizure, scoring, etc. One of the major
difficulties in predicting failure in the case of load-affected resistance is associated with
the need for the history of the applied load. For example, the rate of wear of car brakes
depends on surface conditions, materials of the mating parts, and geometry, as well as
on the history of the applied forces and speeds of rotation.

In reliability engineering it is important to distinguish between deterministic
loads, when the magnitude of the load can be predicted at any moment in time with

certainty, and random loads, when the load magnitude at any moment in time can be predicted only with some probability. Although this distinction is purely methodological, since both types of load produce a similar effect on the product—deterioration and failure—it is important when determining the mathematical tools and the experimental methods needed to investigate the time to failure.

1-4 RELIABILITY AND PROBABILITY

Reliability is concerned with prediction of the lifetime of a product. As follows from the uncertainties associated with product design, development, and service, the lifetime is a random quantity and can be predicted with only some degree of confidence. This means that it is impossible to make a prediction for any specific representative of the population; instead, new measures characterizing the entire population are needed. These new measures, represented by statistics, are supposed to be unchanging for a given population. Statistics, in this context, is a data set of lifetimes and can be represented in the form of a relative frequency histogram, which shows a number of products with respect to the size of the population that has failed in a given time interval. A relative frequency diagram is an experimental set of data and, in principle, contains full information about product reliability; all other statistical measures can be derived from it. A mathematical model of relative frequency is a probability. More correctly, it can be stated that if the size of the population tends to infinity, then the relative frequency for an infinitesimally small lifetime interval tends to a constant value, and a histogram, which is a piecewise constant function, is transformed into a continuous function, called a failure density function. In the theory of probability this function is called a probability density function. In practice, the true failure density function is never known for the simple reason that the size of the population under observation is always finite. This transition from a set of limited experimental data contained in a histogram to a mathematical model represented by the reliability density function introduces a degree of uncertainty that belongs to the cognitive type of uncertainties and will be discussed in detail later. At the moment, however, it is important that in the theory of reliability the failure density function is used as a mathematical model of product reliability.

Reliability is concerned with no-failure performance, and the outcome of any product life observation is either failure or survival. Let us assume that the size of the population is N and that after t years, N_f products have failed, while N_s products have survived, where $N_f + N_s = N$. Then the ratio N_s/N is a measure of probability of the product to operate after t years, that is, it is a measure of reliability of the product at time t. More correctly, if T is the time to failure, which means that T is a random variable, then the probability that this time exceeds any specific given time t, that is, $P(T > t)$, measures the reliability $R(t)$ of the product at time t. Accordingly, the estimation of reliability for a finite population is

$$R^*(t) = P(T > t) = N_s(t)/N \qquad (1\text{-}3)$$

when $t = 0$, $N_s(0) = N$, and it follows from Eq. (1-3) that $R(0) = 1$. At the other extreme, when $t \to \infty$, then $N_s \to 0$, and it means that $R \to 0$. Thus in general, $R(t)$ is a decreasing (or at least nonincreasing) function. Alternatively to Eq. (1-3), the unreliability $F(t)$ is measured by the probability of failure before time t, that is, $P(T \leq t)$, and can be expressed as

$$F^*(t) = P(T \leq t) = N_f(t)/N \qquad (1\text{-}4)$$

Failure and survival are two complementary, that is, mutually exclusive events, and it is obvious that

$$R^*(t) + F^*(t) = 1 \qquad (1\text{-}5)$$

In Eqs. (1-3) and (1-4), $R^*(t)$ and $F^*(t)$ are piecewise constant functions that depend on the size of the population. For an infinitely large population these functions become continuous and unique and are expressed through a failure density function $f(t)$ as

$$R(t) = \int_t^\infty f(\tau)\, d\tau \qquad (1\text{-}6)$$

and

$$F(t) = \int_0^t f(\tau)\, d\tau \qquad (1\text{-}7)$$

Note that Eq. (1-5) holds in the case of continuous functions as well. It should be remembered also that in the theory of probability, $F(t)$ is called a cumulative function.

The foregoing gives the impression that the difference between the theory of reliability and the theory of probability is only in terminology. This is not the case. The theory of reliability uses probability as a mathematical tool to solve engineering problems. Reliability engineering deals with system configurations, testing of products, extending product lifetime, etc., and uses specific measures estimating the state of the engineering system and its trends.

Failure Rate

One of the most important characteristics of product reliability is the rate at which similar products fail as a function of time. If the unreliability function Eq. (1-7) is known, then failure rate is simply a time derivative

$$f(t) = \frac{dF(t)}{dt} \qquad (1\text{-}8)$$

It is seen that the rate of failure in time is represented by the failure density function. As it follows from the relationship similar to Eq. (1-5) for continuous functions, the rate of change of reliability in time is

$$\frac{dR(t)}{dt} = -f(t) \qquad (1\text{-}9)$$

The physical meaning of a failure density function can be better understood if it is looked upon as a limit of the expression

$$f(t) = \lim \frac{1}{N} \frac{\Delta N_f}{\Delta t} \quad \text{when } \Delta t \rightarrow 0 \quad \text{and} \quad N \rightarrow \infty \qquad (1\text{-}10)$$

where N is the size of the population and ΔN_f is the number of products failed over the time interval Δt. As is seen from Eq. (1-10), $f(t)$ estimates, with respect to the whole population, a relative number of components failed over the time interval Δt and is, in effect, a relative frequency histogram. In practice, another indicator of failure rate is very important, namely, with respect to the surviving part of the population, the relative number of components that failed over the time interval Δt. This rate is found from

$$h(t) = \lim \frac{1}{N_s(t)} \frac{\Delta N_f}{\Delta t} \quad \text{when } \Delta t \to 0 \quad \text{and} \quad N_s \to \infty \quad (1\text{-}11)$$

The function $h(t)$ characterizes the trend of the failure rate. If, over the same time interval Δt, the ratio $\Delta N_f/N_s(t)$ remains the same, then although the number of surviving products becomes smaller, the number of components failing over the period Δt also becomes smaller and the trend of the failure rate does not change. If, however, the number of components failing over the time Δt remains the same for any moment t, then for a finite population, $h(t)$ increases. In general, the trend of the failure rate according to Eq. (1-11) is a measure of the tendency to failure as a function of age.

Multiplying and dividing the righthand side of Eq. (1-11) by N, taking into account Eq. (1-10) and the fact that when $N_s \to \infty$ and $N \to \infty$, $\lim N_s (t)/N \to R(t)$, gives an alternative form:

$$h(t) = \frac{f(t)}{R(t)} \quad (1\text{-}12)$$

Another expression relating $h(t)$ and $R(t)$ is obtained if $f(t)$ is substituted in Eq. (1-12) from Eq. (1-9):

$$h(t) = -\frac{1}{R(t)} \frac{dR(t)}{dt} \quad (1\text{-}13)$$

and the latter is integrated, taking into account that $R(0) = 1$:

$$\int_0^t h(\tau) \, d\tau = -\int_1^{R(t)} \frac{dR(\tau)}{R(\tau)} \quad (1\text{-}14)$$

As a result of the integration, yet another relationship between the reliability and failure rate functions is obtained:

$$R(t) = \exp\left[-\int_0^t h(\tau) \, d\tau\right] \quad (1\text{-}15)$$

The function $h(t)$ is called the hazard function, the hazard rate, the mortality rate, or the instantaneous failure rate. In this text, the hazard function term is used as an indication of the state of the product with respect to its survival during the operating life.

The reliability function, the hazard function, and the failure density function are various measures of the same phenomenon: scatter of failure times for a population. Since any of these functions uniquely describes the phenomenon, they are not independent, but interrelated, as is seen from Eqs. (1-9), (1-12), (1-13), and (1-15).

The following example demonstrates the application of concepts introduced in

Table 1-1 Grouped Failure Data for dc Electric Motors

Interval in days	200–300	300–400	400–500	500–600	600–700	700–800	800–900	900–1000
Number of failures	1	4	20	40	50	15	7	1

this part. In the example a population of a finite size is considered, which means that all found measures should be considered as estimates.

Example 1-8 A total of 138 dc electrical motors were observed, and the time to failure for each of the motors was recorded. It was found that no motor survived beyond 1000 days of operation. The 1000 days then was divided into 10 subintervals, and a table of records of all failures that occurred within each subinterval was made (Table 1-1).

The results in Table 1-1 in graphical form would represent a frequency of failures histogram. If, however, the number of failures in each interval is divided by the overall number of motors under observation (138), then a relative frequency histogram can be plotted (see Fig. 1-5).

Note that the number of subintervals is arbitrary. In choosing this number, it should be kept in mind that a large number of subintervals results in insignificant information for some of them, whereas a small number distorts the distribution of failure times. The number of subintervals depends on the number of products under observation. Usually from 10 to 15 subintervals are sufficient in applications. However, it can be fewer in cases of limited data.

The relative frequency histogram is an estimation of the failure density function. If that function could be found, for example, by assuming some mathematical model of failure distribution, then Eqs. (1-6), (1-7), and (1-12) could be used to determine the reliability, unreliability, and hazard functions. An alternative is to estimate the reliability and hazard functions by piecewise functions using the data from the relative frequency histogram. The drawback of utilizing the piecewise functions is that they are given either in graphical or tabulated form that is not convenient when further analysis of reliability involving the interrelationships among the various components is required. It should be pointed out here that for a finite population (or sample) the estimated statistics are random, which means that for another set of failure times of similar dc motors, the relative frequency histogram may be different.

In the following it is shown how to use empirical data to get estimates of reliability, failure density, and hazard functions.

To plot the above-mentioned functions, the following steps should be taken:

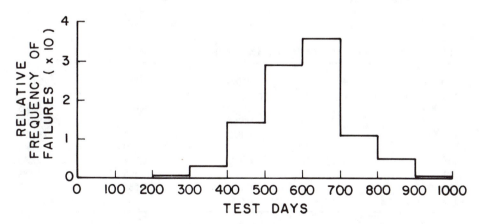

Figure 1-5 Histogram for the data in Table 1-1.

1. Divide a data axis (time axis) into intervals (usually of equal length) and count the number of failures in each interval, that is, transform the raw data into grouped data. The data in Table 1-1 are an example of grouped data.
2. Calculate the estimates of reliability values at the midpoint of each interval i by using Eq. (1-3):

$$R^*(t_i) = \frac{N_s(t_i)}{N} \tag{1-16}$$

where $N_s(t_i)$ is the number of units surviving at time t_i and N is the total number of units under observation. Note that t_i is the midpoint of the ith interval.
3. Calculate the estimates of failure density values for each interval by using a discretized version of Eq. (1-9):

$$f^*(t_i) = -\frac{R^*(t_i) - R^*t_{i-1})}{\Delta t_i} \tag{1-17}$$

where Δt_i is the length of the ith interval, $\Delta t_i = t_i - t_{i-1}$.
4. Calculate the estimates of the hazard function values by using a discretized version of Eq. (1-12):

$$h^*(t_i) = \frac{f^*(t_i)}{R^*(t_i)} \tag{1-18}$$

Note that for a small sample size, instead of Eq. (1-16), the following should be used:

$$R^*(t_i) = \frac{N_s(t_i) + 0.7}{N + 0.4} \tag{1-19}$$

The rationale for this estimator is that $i = n$ means that $R^*(t_n) = 0$ if all units have failed by this time. This situation is unlikely if sample size is increased. Formula Eq. (1-19) corrects this discrepancy.

The computations for the dc motor test data are shown in Table 1-2.
The plots of $R^*(t_i)$, $f^*(t_i)$, and $h^*(t_i)$ are shown in Figs. 1-6, 1-7, and 1-8, respectively.
It should be noted that the relative frequency histogram (Fig. 1-5) and the failure density function (Fig. 1-7) represent the same phenomenon. However, a slight difference is possible when Eq. (1-19) for $R^*(t_i)$ is used.
Graphical representation of reliability and hazard functions is a convenient way of assessing the performance of the product. It is seen from Fig. 1-6 that the reliability of dc motors starts dropping sharply somewhere between 400 and 500 days of operation. After this period of operation, the hazard function goes up at an increased rate. This information can be used, for example, to set up a preventive maintenance schedule or to make a decision on the number of spare parts or back-up products needed.

Table 1-2 Computations for the dc Test Data

Interval[†]	$N_s(t_i)$	$R^*(t_i)$	$f^*(t_i)10^2$	$h^*(t_i)10^2$
200–300	137	0.995[‡]	0.005	0.005
300–400	133	0.966	0.029	0.030
400–500	113	0.822	0.144	0.175
500–600	73	0.533	0.289	0.542
600–700	23	0.172	0.361	2.099
700–800	8	0.064	0.108	1.688
800–900	1	0.013	0.051	3.923
900–1000	0	0.006	0.007	1.166

[†]Here, $200 \le t_i < 300$.
[‡]In estimating $R^*(t_i)$, Eq. (1-19) is used where $R^*(t_0) = 1$.

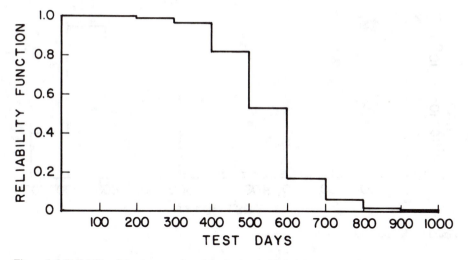

Figure 1-6 Reliability of the dc motors based on the data in Table 1-1.

Mean Time to Failure

The mean time to failure (MTTF) for a product is equivalent to the mean of the probability distribution of times to failure and is found by averaging the failure density function,

$$MTTF = \int_0^\infty \tau f(\tau)\, d\tau \qquad (1\text{-}20)$$

A relationship between the reliability function and the MTTF is easily obtained by substituting $f(t)$ from Eq. (1-9):

$$MTTF = -\int_0^\infty \tau\, dR(\tau) \qquad (1\text{-}21)$$

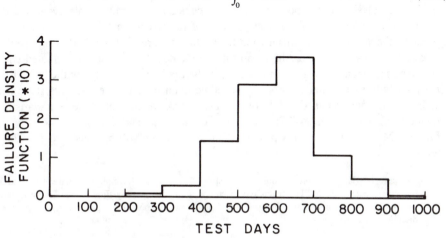

Figure 1-7 Failure density function for the dc motors.

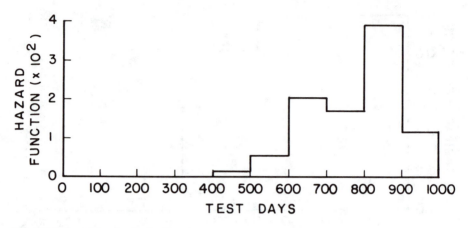

Figure 1-8 Hazard function for the dc motors.

Integrating the latter by parts,

$$\text{MTTF} = -tR(t)\,\big|_0^\infty + \int_0^\infty R(\tau)\,d\tau \tag{1-22}$$

and taking into account that $R(0) = 1$ and $R(\infty) = 0$, the MTTF becomes

$$\text{MTTF} = \int_0^\infty R(\tau)\,d\tau \tag{1-23}$$

An alternative expression giving a relationship between the MTTF and the hazard function is found by substituting Eq. (1-15) into Eq. (1-23):

$$\text{MTTF} = \int_0^\infty \exp\left[-\int_0^t h(\tau)\,d\tau \right] dt \tag{1-24}$$

The notion of MTTF can be applied to both simple components and complex products. In the case of simple components, a failure constitutes an irreversible event and the MTTF is an estimation of the average time of this event. In the case of a complex product, the MTTF is the mean time to the first failure. If the failed component is repaired or substituted, then the mean time to the next failure can be estimated when another component fails. For complex products it is important to know the cycle time between failures in order to properly schedule maintenance times. At this stage it is sufficient to note that in the case of a constant hazard function there is a statistical relationship between the MTTFs of simple components and the mean time between failures (MTBF) of a complex product. This relationship will be considered in Chapter 4.

Example 1-9 The data obtained during the test of dc motors (see Example 1-8) make it possible to find an estimate of the MTTF for this type of motor. To do this, any equation for MTTF given above can be used if summation is substituted for analytical integration. For example, Eq. (1-23) for a discrete data set becomes

$$\text{MTTF}^* = \sum_{i=1}^{n} R_i^*(t_i) \cdot \Delta t_i \tag{1-25}$$

Formula Eq. (1-25) expresses the area under the piecewise function in Fig. 1-6. For the case under consideration, MTTF* = 557 days, which is when the hazard function starts going up sharply (see Fig. 1-8).

PROBLEMS

1-1 What are the objectives of reliability and maintainability engineering and how are the two related?

1-2 A pressure vessel consists of two parts connected by bolts (see Fig. P1–2). The tightening of the bolts should be such that no leakage occurs. Identify physical uncertainties that may affect the possibility of leakage.

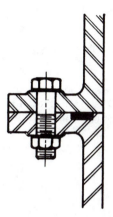

Figure P1-2

1-3 Identify the physical uncertainties in a brake shoe (see Fig. P1–3) that may affect the braking torque. Assume that the shoes are worn in.

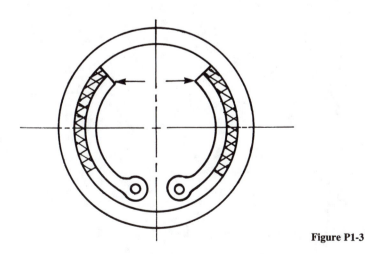

Figure P1-3

1-4 In similar ships, 10 failures of propulsion systems have been reported. What information would you request with respect to these ships in order to assess the reliability of the propulsion system?

1-5 Mr. Jones asked Mr. Smith how reliable his new car was. Mr. Smith replied that it was 99% reliable. Is this answer correct?

1-6 What is the difference between the failure rate function $f(t)$ and hazard function $h(t)$? If they are related, then why is there a need for an independently defined hazard function?

1-7 A reliability function is a function nonincreasing in time. The following are three decreasing functions that can be taken as mathematical models of reliability:

$$R_1(t) = \left(\frac{1}{1 + \lambda t}\right)^{1/\lambda}$$

$$R_2(t) = \exp\left(-\frac{t}{\lambda}\right)$$

$$R_3(t) = \exp\left[-(t + \frac{1}{3}\lambda^2 t^3)\right]$$

(a) What are the hazard functions for the above models? Plot these functions for the case $\lambda = 1$.
(b) What are the qualitative differences between the three hazard functions?
(c) Do you think that all or any of these functions may reflect reality?

1-8 The following test results represent the kilocycles to failure of 30 aluminum specimens: 80, 98, 103, 104, 108, 112, 116, 119, 125, 130, 131, 134, 135, 139, 144, 145, 149, 151, 152, 153, 157, 160, 162, 164, 166, 167, 168, 170, 174, and 180. Compare hazard functions for three different time intervals: $\Delta t = 5, 10,$ and 15 kilocycles. What time interval seems most suitable?

CHAPTER TWO

INTEGRATING RELIABILITY INTO DESIGN

2-1 INTRODUCTION

The process of integrating reliability into design starts at the conceptual stage and ends at the detailed design stage. From a concept to working drawings, information about the product grows during the design process. Reliability considerations affect the trend of product development, and at the same time, product development widens the data base for reliability assessment of the product. In effect, this is a process with constant feedback that tends to optimize, consciously or unconsciously, the design. Thus the process of integrating reliability into design can be seen as an interative process. As such, it may comprise a different number of steps in each specific case. However, what is important is to identify qualitatively different steps, since they entail qualitatively different methods of reliability assessment. If a relation between different methods and different design stages is found, then for each design stage, clear reliability objectives can be formulated. This is important, since objectives, methods of analysis, and available data base should be in agreement.

A product is a physical system comprising interacting components. Each specific arrangement of components into a physical system constitutes a configuration. Any concept entails a specific configuration. For example, the concept of an internal combustion engine can be outlined by sketching a crankshaft, connecting rod, piston, and cylinder. A six-cylinder engine would have a configuration different from a four-cylinder one. However, an internal combustion engine is more than a crankshaft-connecting rod-piston-cylinder system. It includes a fuel system, lubrication system, control system, etc. It is clear that during the design process the system grows in terms of a number of components and interfaces between components. The reliability of a physical system can be found as long as reliability of components and interfaces, and a method of estimating reliability of a system based on reliability of components and their interfaces, is known. Thus, if there was a method of estimating system reliability, then the different objectives at different design stages would only be due to the different design data bases. Conceptual, preliminary, and detailed design stages have qualitatively different information about the design and thus lead to three different stages of reliability analysis. However, the common ground for all three stages is the concept of system reliability.

2-2 SYSTEM RELIABILITY

Let us consider a shaft assembled with two bearings in a housing (Fig. 2-1) that is a part of a larger system, say, a transmission box. Each component in this assembly performs a specific role: bearings reduce friction losses and maintain shaft alignment and position, seals prevent oil leakage out of the housing, etc. The question that each design engineer asks is, "what may prevent a component from performing its function as anticipated?" In other words, "what affects the reliability of each component?" The next question is, "how does the performance of a specific component affect the performance of the entire system?" Finding answers to these questions is what reliability engineering is all about.

Let us consider the lip-type sealing in Fig. 2-1. Its performance depends on properties of the material it is made of, oil temperature, oil pressure, initial pre-tension of the ring, speed of shaft rotation, shaft vibration, and fit between the shaft and seal, to name a few factors. What is important is that some of these factors such as seal material and initial pre-tension are associated with the seal itself, whereas other factors such as oil temperature, oil pressure, and fit between the seal and shaft are associated with the fact that the seal is part of the system. The latter factors can also be broken down into two categories: some are due to seal-system interaction (like contact pressure between the seal and the shaft), while others are due to the common system environment (like shaft vibrations, temperature of oil). What is apparent from this example is that since reliability of the seal depends on all factors, it has meaning only in a system, since it reveals itself only when the seal is functioning. What is also clear is that the same seal might have different reliability characteristics in different systems. Simply speaking, a good seal may perform well in one design and may fail, if badly installed or subjected to strenuous temperatures and stresses, in another design.

A product is an assembly of interacting components. The role of a component in

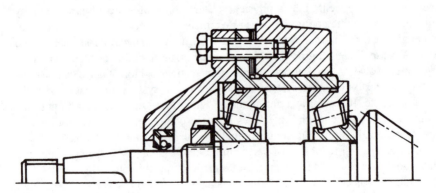

Figure 2-1 Bevel-gear-shaft assembly.

an assembly is twofold: it affects system performance and, at the same time, is itself affected by the system. The effect of a component on a system depends on that component's own properties (material, geometry, surface conditions, etc.) and on the position of this component with respect to the others as a functional element (sharing load with other components, redundant, standby). The effect of the system on a component depends on the interfaces with mating components (fits, contact conditions, compatability of materials), on the internal environment, which is the result of product functioning (temperature, vibration, humidity, etc.), and on the external environment.

Usually the component's own properties are kept within specified limits by the quality control system. The interfaces between the components are less controllable, since interfaces are the result of system assembly and in this respect the only verification possible is to operate the product. Note that if components' specifications are met, then the limits of some interfaces (fits, clearances, surface hardness) are known, whereas the limits of others (like friction) remain uncertain. The greatest uncertainty is associated with the internal environment in the functioning product, first, because of the complexity of interrelated processes and, second, because of the influence of the external environment (temperature, humidity, dust, etc.) on the internal environment.

The prediction of reliability of a product depends on the ability to differentiate all possible causes of failure and to weigh their potential impact. Sometimes mere indentification of causes and their potential impact may be helpful. In Fig. 2-1 a groove in a shaft acts as a stress concentrator and may cause shaft failure; an interface between the lip-seal and the shaft may be too tight and may lead either to a groove in the shaft, if the shaft surface is not hard enough, or to lip wear; or overheated oil may cause failure of a bearing and seal. However, in reliability the indentification of causes is only the first step. The next step is to find out the quantitative effect of each cause on component and system reliability. In this respect it is important to know what information is needed to make such an assessment; once the type and amount of information needed is identified, then it becomes clear at what stage of the design process the reliability of the product can be estimated.

Let us consider the shaft in Fig. 2-1 again. What information is required to determine its reliability? In broad categories the information concerns the functional role of the shaft, the shaft geometry and material properties, external loads, interaction with other components, and internal environment in the gearbox. Only the functional role of the shaft is known at the conceptual stage, more information becomes known at the preliminary stage (shaft geometry, material properties, external loads, and to some extent, internal environment), and even more specific information is available at the detailed design stage (when surface finish and hardness are finalized, interaction with other components due to assigned tolerances can be assessed, and type of lubrication decided). However, even at the final design stage there is not enough information to predict the scatter of shaft failures in time caused by fatigue. This scatter depends on the scatter of fatigue properties of the material itself, on the effect of various additional factors on fatigue (such as surface finish, shaft size, stress concentration, temperature), and on the effect of the type of stresses (symmetrically reversed, nonsymmetrically reversed, complexity of the stress state) and their magnitude. It should be made clear that most of this information is not available not only because it would be very expensive and time consuming to obtain it, but also because it is impossible to simulate exactly the external loads and environments and the effect of other components at the interfaces when a shaft is tested.

The example of a shaft creates a pessimistic impression and may lead to a wrong conclusion that since the true reliability level cannot be estimated at the detailed design stage, it should be postponed until the service stage, when the product's life data become available. The pessimism would be justified to some extent if there were no previous experience, knowledge, or similar designs. It is exactly for these reasons that the application of reliability methodology to new designs increases the level of confidence in this design. Experience is associated with the individuals involved in the design process, with their personal knowledge and range of interests. Knowledge is a state of science in a specific field and should not be subject to interpretation, i.e., it should be specific and complete in terms of conditions, limitations, and results. And finally, similar designs serve as reference points in reliability predictions, a point that merits some elaboration.

The question of reliability of products that are similar in design and functioning but have different input-output parameters is important in mechanical reliability. It comes down to a question whether two similar designs have the same reliability characteristics. Turbines, engines, compressors, pumps, boilers, etc., often have similar designs but different power outputs.

In general, two similar designs do not have the same reliability properties because they do not belong to the same population of products, which means that there may be no similarity in such properties as their resistance to fatigue, wear, and corrosion, manufacturing and quality control (if, for example, a geometrically similar part is forged instead of machined, or operational characteristics are different due to different power parameters of two products). However, in practice, proven design solutions are always adopted for a larger or a more powerful machine. Although use of the smaller prototype does not guarantee the actual level of reliability, it reduces uncertainty and speeds up development of the finished product.

Since the information needed to assess product reliability becomes available after the design is finished and from the results of operational experience, at what stage does the design engineer then become concerned with the reliability aspect of product development? In this respect, it is important to understand that the objective of the design process, as far as reliability is concerned, is to increase the relative level of reliability irrespective of its absolute level, which in most cases remains unknown. The methods of increasing the relative level of reliability are very well developed and understood. These methods are different at different design stages because of changing design objectives and widening of the data base.

2-3 CONCEPTUAL DESIGN STAGE

The information available at the conceptual design stage is the configuration of the product or, more correctly, various configurations of competing alternatives. Then the natural question is whether it is possible to compare various alternatives from the point of view of their reliabilities. If the answer is positive, reliability could be used, among other criteria, for choosing an alternative.

An alternative does not necessarily mean another arrangement of the same components but may include new components or a different number of components. The reliabilities of components may be very sketchy at this stage; the reliability of interfaces between components may be even less definite. It is clear that a comparative assessment of various configurations can be based on specific assumptions common to all configurations, thus providing a common ground for all configurations.

Any configuration is a result of some functionally required arrangement of components or subsystems. In some configurations, all components operate as long as the system is operating, whereas in others some components or subsystems are in a standby position ready to be switched in (automatically or manually) as soon as an identical part fails. In those systems where all components operate, there may be redundant components, so that if one fails, the system continues to function. These redundant components may or may not be affected by the failure of an identical component. Thus the arrangement of components in a configuration entails its specific functional properties that are a characteristic of the configuration itself. Since reliability is concerned with the integrity of functional properties, it follows that each configuration can be characterized by its reliability properties. So, in principle, various configurations can be compared on the basis of their reliability properties. There cannot be any reliability of the scheme, since reliability is a property of a physical system. So two schemes can be compared only if some properties of components in both schemes are assumed. However, as mentioned earlier, properties of components are vague at the conceptual stage, and thus the results of any reliability comparison have a qualitative rather than a quantitative nature. They should be viewed with the sense that one configuration has a higher potential for reliability growth than the other. This point will be considered further in the following sections.

Reliability Block Diagram

Reliability is associated with satisfactory performance. When analyzing a physical system comprising interacting components from the point of view of reliability, it is important to identify the effect of each component on system reliability/performance, specifically, from the point of view of whether component failure results in system failure. In a chain, segment failure results in chain failure; similarly, bearing failure may result in transmission failure. A segment in a chain and a bearing in a transmission operationally play similar roles, and by extending the analogy, a bearing, like a segment, is functionally in series with all other components in a transmission. In general, *a component in a system is functionally in series to all other components if its failure results in system failure.* If, however, there are redundant components, so that a system may fail only if all redundant components fail, then these components are considered to be functionally parallel to each other in a system.

The role of components in system performance/reliability can graphically be shown on a diagram in which each component is represented by a block. Such a diagram is called a functional diagram or reliability block diagram. An interface between two components can also be represented on this diagram by a separate block if failure of this interface is a result of interaction between two components and cannot be associated with either of two components taken on their own. A press-fitted gear on a shaft is an example of such an interface, since failure of a press-fit is the result of gear-shaft interface conditions. The role of an interface in a block diagram is similar to that of any component, namely, if its failure results in system failure, it should be in series to all other blocks; otherwise it is in parallel. It is important to note that the reliability block diagram gives no indication of the effect of the system on a component.

The following well-known example illustrates the difference between the arrangement of components in a physical system and their representation in a reliability block diagram.

Example 2-1 Consider a schematic physical diagram of two valves in a pipeline (Fig. 2-2a). Would the reliability block diagram of this system appear as a series or as a parallel arrangement of blocks? The proper answer depends on the definition of the adequate performance of the system. If the two valves are normally shut but are expected to open on command to provide flow, then this is a series system in terms of reliability (Fig. 2-2b), since if any one of the valves fails, the whole system fails. If, however, the two valves are normally open but are expected to shut on command to stop flow, then this is a parallel system in terms of reliability (Fig. 2-2c), since if any one of the valves fails, the other would close the pipe, so that the system does not fail.

Example 2-1 illustrates that as long as the functional role of any component or interface is understood, the reliability block diagram can be drawn for that system. A reliability block diagram does not indicate, however, any level of reliability. Reliability can be found if the diagram is supplemented with information about reliabilities of components and interfaces.

In general, a reliability block diagram comprises some blocks in series and some in parallel. Before considering this general case, two particular situations will be discussed, namely, when all components are either in series or in parallel.

Components in series. Let us consider a specific system comprising a boiler, a turbine, and pipes connecting them. It is clear that three blocks representing three elements of this system are in series to each other on the reliability block diagram (see Fig. 2-3).

The definition of reliability does not discriminate between a component or a system. Thus if T is the time to system failure, then the probability that it exceeds any given time t defines the reliability of the system. For the system in Fig. 2-3, reliability according to this definition is

$$R_s(t) = P[E_s(T > t)] \tag{2-1}$$

where $E_s(T > t)$ is the no-failure event. For this system the no-failure event holds only if the no-failure event is true for the boiler, piping, and turbine simultaneously,

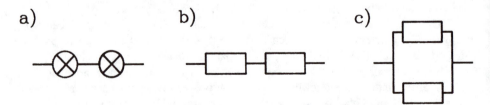

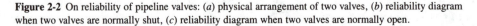

Figure 2-2 On reliability of pipeline valves: (*a*) physical arrangement of two valves, (*b*) reliability diagram when two valves are normally shut, (*c*) reliability diagram when two valves are normally open.

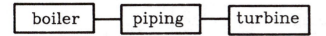

Figure 2-3 Reliability diagram of a system with components in series.

i.e., that the time to failure T exceeds any given time t for all three components. This event is mathematically represented as a logical multiplication in the form

$$E_s (T > t) = E_b (T > t) \cap E_p (T > t) \cap E_t (T > t) \qquad (2\text{-}2)$$

which means that in order for event E_s to be true, all three events E_b, E_p, and E_t must be true. Thus reliability of the system according to Eq. (2-1) is

$$R_s (t) = P (E_b \cap E_p \cap E_t) \qquad (2\text{-}3)$$

If the state of each of three components does not affect the state of any other three components [e.g., corrosion of pipes does not affect the state of the turbine, or cracks in the boiler do not affect the vibration (and thus the fatigue) of pipes], or in general, if there are no common-cause failures, then the three events E_b, E_p, and E_t can be considered statistically independent. The assumption of statistical independence should always be questioned, however. For example, if pipes and a turbine are rigidly connected, then they constitute a jointly vibrating system and may influence each other's failure rate. The question of statistical dependency of random events does not have a simple yes or no answer, since there may be various degrees of dependence among them. The effect of statistical dependency will be considered further. At the moment, let us assume that in Eq. (2-3) all three events are statistically independent. The probability that all three events are true equals a product of probabilities that each of them is true, i.e.,

$$R_s (t) = P (E_b) P (E_p) P (E_t) \qquad (2\text{-}4)$$

and Eq. (2-4) is equivalent to

$$R_s (t) = R_b (t) R_p (t) R_t (t) \qquad (2\text{-}5)$$

In general, for n components in series, in the case when they do not influence the failure rate of each other, the reliability of the system $R_s(t)$ is found as a product of reliabilities of components $R_i(t)$:

$$R_s(t) = \prod_{i=1}^{n} R_i (t) \qquad (2\text{-}6)$$

where $\prod$ denotes the product sign. Thus the reliability of a system represented by a functional diagram of its components arranged in series is the product of reliabilities of its components. Any nonredundant system is represented by a series block diagram.

Components in parallel. Let us consider now a system in which a boiler supplies steam to two turbines. Since failure of one of two turbines or connecting pipes does not result in the failure of another piping-turbine subsystem, the two subsystems must be functionally parallel to each other (Fig. 2-4). The question is how to find the reliability of a system comprising two parallel subsystems (Fig. 2-5). It is clear that for

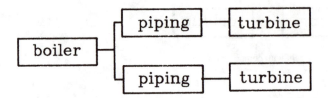

Figure 2-4 Reliability diagram of a system with components in series and parallel.

the entire system to fail, two subsystems must fail. Since the event of failure represents the unreliability by definition, it follows that the unreliability of the system in Fig. 2-5 equals the probability of failure of two subsystems:

$$F_s(t) = P[\bar{E}_1(T < t) \cap \bar{E}_2(T < t)] \tag{2-7}$$

where $\bar{E}_1$, $\bar{E}_2$ denote the failure events for two subsystems. Again, if two subsystems in Fig. 2-5 do not affect each other, the two events of failure in Eq. (2-7) can be considered statistically independent. The corresponding formula for system unreliability becomes

$$F_s(t) = P(\bar{E}_1)P(\bar{E}_2) = F_1(t)\,F_2(t) \tag{2-8}$$

In general, for n statistically independent components in parallel the unreliability (failure) function $F_s(t)$ is expressed as a product of unreliabilities of components $F_i(t)$:

$$F_s(t) = \prod_{i=1}^{n} F_i(t) \tag{2-9}$$

Any fully redundant system is represented by a parallel block diagram. Recalling that $F(t) = 1 - R(t)$, Eq. (2-9) can be written in terms of reliabilities as follows:

$$R_s(t) = 1 - \prod_{i=1}^{n} [1 - R_i(t)] \tag{2-10}$$

Components in series and parallel. Systems with partial redundancy are represented by a block diagram in which some blocks are in series while others are in parallel. This situation is shown in Fig. 2-4. Formal representation of the reliability of the system is based on the application of principles already known. The method of system reduction can be used to find the system reliability. For example, in the case of Fig. 2-4, first the reliability of a piping-turbine subsystem would be found using Eq. (2-6).

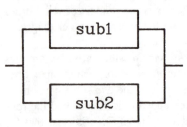

Figure 2-5 Reliability diagram of two subsystems in parallel.

Then the reliability of two subsystems in parallel would be found using Eq. (2-10). And last, the system reliability would be found by multiplying the latter by boiler reliability, since the two would be in series.

On Comparison of Reliability Block Diagrams

Let us assume that two competing configurations are represented by two block diagrams that differ in terms of the number of components and their functional role. If the reliability functions for each component in the two diagrams were known, then formulas in the previous section would reveal reliabilities of two configurations at any moment in time as long as failures of components are statistically independent and functionally similar components have the same reliability functions. The following example illustrates the situation.

Let us consider again two configurations of a power unit represented in Figs. 2-3 and 2-4. The reliability function of the configuration in Fig. 2-3 is given by Eq. (2-5):

$$R_{s1}(t) = R_{b1}(t)R_{p1}(t)R_{t1}(t) \tag{2-11}$$

and for the configuration in Fig. 2-4 by

$$R_{s2}(t) = R_{b2}(t) \{1 - [1 - R_{p2}(t)R_{t2}(t)]^2\} \tag{2-12}$$

where subscripts 1 and 2 denote configurations according to Figs. 2-3 and 2-4, respectively.

In general, in mechanical systems the reliabilities are not the same for functionally similar but geometrically different components or subsystems. Also reliability functions in two configurations will differ if the replacement of failed parts is taking place. However, if it can be assumed that $R_{p1}(t) = R_{p2}(t) = R_p(t)$ and $R_{t1}(t) = R_{t2}(t) = R_t(t)$, then it follows from Eq. (2-12) that for the term in square brackets the following inequality holds for any moment in time, except $t = 0$,

$$R_p(t)R_t(t)[2 - R_p(t)R_t(t)] \geq R_p(t)R_t(t) \tag{2-13}$$

because the term in brackets is larger than unity, except at $t = 0$. The term on the righthand side of Eq. (2-13) corresponds to the reliability of the pipe-turbine subsystem in Fig. 2-3, and the term on the lefthand side, to Fig. 2-4. Thus, if the reliability functions of pipes and turbines are the same in both schemes, then the reliability of the configuration with parallel arrangement of components in Fig. 2-4 is higher than that in Fig. 2-3 at any moment in time. In general, if functionally similar components have the same reliabilities, then their parallel arrangement increases the reliability of the system.

In Eq. (2-13) the term $[2 - R_p(t)R_t(t)]$ is the factor of increase in reliability and does not remain constant but is increasing in time from 1 at $t = 0$ to 2 at $t \rightarrow \infty$. This fact indicates that the advantage of redundancy is time-dependent and thus the comparison of different configurations in mechanical systems should be done with an indication of time, i.e., the probabilities that two systems will survive a specified time.

At the conceptual stage, reliability functions of components in various alternative configurations most often are not known. However, since reliability is of secondary importance at this stage, various alternatives can be compared in relative terms by assuming some levels of reliability and keeping them the same for similar components in all configurations, which is equivalent to assuming some point in time for which a comparison is made. Then, formally, time is excluded from consideration in this case. It should be understood that the results of such an analysis have a qualitative nature and that this analysis is valid for statistically independent components that are in a state of probabilistic similarity. When the assumption of a common time base is not valid, then the assessment of reliability changes. Let us assume that in Fig. 2-3 at some point in time t_{i1} a component in the turbine failed and a spare part was installed. Then the reliability of the system after repair, i.e., for any time $t \geq t_{i1}$, will be

$$R_{s1}(t) = R_{b1}(t) \, R_{p1}(t) R_{t1}(t - t_{i1}) \tag{2-14}$$

where reliability functions for the boiler and pipes are in an old time scale and reliability for the turbine is in a new time scale.

Let us assume, for the sake of simplicity, that in an alternative configuration (Fig. 2-4) one of two turbines has also failed at time $t = t_{i2}$. Then the reliability of the system after repair, i.e., for any time $t \geq t_{i2}$, will be

$$R_{s2}(t) = R_{b2}(t)\{1 - [1 - R_{p2}(t)R_{t2}(t)][1 - R_{p2}(t)R_{t2}(t - t_{i2})]\} \tag{2-15}$$

where again the time base for one of the turbines has changed. The comparison of two configurations is not straightforward anymore because the two times of failures t_{i1} and t_{i2} are random and affect the statistics of failures. Below, the Markov method and, in Chapter 5, a simulation technique allowing assessment of the reliability of systems with repair, will be considered.

In general, when comparing various alternatives, three assumptions are very important: common time base, statistical similarity, and statistical independence of failures. A brief discussion of the applicability of these assumptions follows.

Time base. Since reliability is a function of time, it is important to identify the reference point in time for all components in a system. In a system the time base for various components is not always the same. For example, in a car transmission box some components operate only when the first gear is engaged, while others operate only when the second gear is engaged, etc. So the total time of transmission operation will consist of a sum of nonoverlapping times of operation at various regimes. These different time bases for different components are important in assessing the system reliability based on reliabilities of its comprising components. The difference in time bases is also relevant to systems with repairs, shared loads, and a standby redundancy.

Statistical similarity. The most common assumption in mechanical reliability is that functionally similar components have the same reliability functions. In general this is not true. Even identical components do not necessarily have similar reliability functions, as was discussed earlier. In this sense, a 100- and a 50-MW turbine cannot be treated as similar products. Usually a larger scale product is designed using an existing smaller scale prototype. It is clear that as far as functional properties of two products are concerned, such as kinematics, efficiency, or output, the prototype allows direct application of similarity laws. However, at the moment there are no explicit similarity laws as far as reliability of two products is concerned. The true reliability of a larger turbine can be found only in service.

Statistical dependence. Components can affect each other's failures in a way not obvious and not reflected in a reliability block diagram in the form of functional relationships. For example, in Fig. 2-1, overheated oil may cause failure of a bearing and a seal. This type of relationship can be discovered if failure records are analyzed statistically. Statistical dependence is an indication of a common cause for failures. So the interdependence between components in a system manifests itself in a deterministic functional relationship, which involves a direct effect of one component on another, and in a statistical dependence, which is a statistically discovered form of functional relationships between components in a system. The reliability block diagram represents only the deterministic functional relationship between the components.

Example 2-2 This example illustrates the effect of statistical dependence on system reliability. In this case the reliabilities of two bearings shown in Fig. 2-1 are considered.

Let us analyze the following life data records: out of 500 transmissions in operation for 1000 hours, 100 transmissions have failed. In 60 cases the left bearings failed (near the threaded part of the shaft) and in 50 cases the right bearings failed. In 10 of these cases both bearings failed at the same time. Note that any bearing failure would be reported as a system failure. These statistical results can be shown graphically by means of the Venn diagram (see Fig. 2-6), where L is the set of all left bearing failures, R is the set of all right bearing failures, and L ∩ R is the set of both bearing failures. Assuming that the frequency of failure equals the probability of failure, the reliability of two bearings after 1000 hours of operation can now be estimated. Let us denote by $\bar{E}_1$ the event of the left bearing failure and by $\bar{E}_2$ the event of the right bearing failure. Then the probability that either of two bearings fails equals

$$P(\bar{E}_1 \cup \bar{E}_2) = P(\bar{E}_1) + P(\bar{E}_2) - P(\bar{E}_1 \cap \bar{E}_2) \tag{2-16}$$

where $P(\bar{E}_1)$ and $P(\bar{E}_2)$ are the probability of failure of the left and right bearings, respectively, and $P(\bar{E}_1 \cap \bar{E}_2)$ is the probability of two bearings failing simultaneously. Since, according to Eq. (1-4),

$$P(\bar{E}_1) = \frac{60}{500} \qquad P(\bar{E}_2) = \frac{50}{500} \qquad P(\bar{E}_1 \cap \bar{E}_2) = \frac{10}{500}$$

the probability of one bearing's failure (unreliability) after 1000 hours equals $P(\bar{E}_1 \cup \bar{E}_2) = 100/500 = 0.2$, and the corresponding reliability is $R(1000) = 1 - 0.2 = 0.8$. However, if the same number of bearings (110) failed but none of these failures took place simultaneously, then $P(\bar{E}_1 \cap \bar{E}_2) = 0$ in Eq. (2-16), and the unreliability is $P(\bar{E}_1 \cup \bar{E}_2) = 110/500 = 0.22$, while the corresponding reliability is $P(1000) = 1 - 0.22 = 0.78$. If, at the other extreme, $\bar{E}_1 \cap \bar{E}_2 = 50$ in Fig. 2-3, i.e., an indication of a full statistical dependency, then

$$P(\bar{E}_1 \cup \bar{E}_2) = \frac{50}{500} + \frac{50}{500} - \frac{50}{500} = 0.10$$

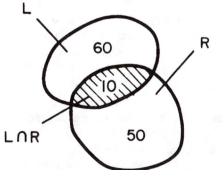

Figure 2-6 Venn diagram of two bearings failures.

and the reliability would be $R(1000) = 0.90$.

To summarize, if the number of failures in the system of two components is the same, as far as each component is concerned, but some failures take place at the same time, then the reliability of the system is higher than it would have been for the same total number of independent failures. This indicates the role of statistical dependence in system reliability, namely, by taking statistical dependence into account, the reliability is increased. In the given example, reliability increased from 0.78 for statistically independent failures to 0.80 for partially dependent failures and to 0.90 for fully dependent failures.

It follows from Example 2-2 that the assumption of statistical independence gives a conservative estimate of system reliability. Since the degree of statistical dependence is rarely known, an assumption of independence is most often used in practice. However, there is another type of statistical dependence, as in the case of shared loads, which does not fall into the above category and should be taken into account. This will be discussed further.

Redundancy in System Design

A parallel arrangement of components in a reliability block diagram represents a system with redundancy. The point is that there are many approaches to incorporating the concept of redundancy in system design. There could be various degrees of redundancy and various levels of redundancy. The notion of degrees of redundancy is related to the number of similar components, units, or subsystems that are arranged functionally in parallel. The notion of levels of redundancy is related to the type of components, units, or subsystems that are arranged in parallel. These two notions are describing different types of redundancy, and their effect on reliability is different as well.

Consider various configurations with various degrees of redundancy. For clarity, assume that a diesel-gearbox-generator unit (see Fig. 2-7a) is used in a remote area as a source of electric energy. Also assume that the unit's reliability function does not depend on the load (power generated) and that the probability that one unit will operate fail-free during one year is 0.95. Then the probability (reliability) of fail-free operation during the same time of two less powerful units in parallel, according to Eq. (2-10), is 0.9975, of three units is 0.999875, and of four units is 0.9999937. A different number of diesel-generator units working in parallel represents different degrees of redundancy. It is clear that with the increase of degree of redundancy the reliability of the system increases. The question is how much, in relative terms, is gained in reliability with the addition of each new degree of redundancy. For this specific example the first redundant unit increases system reliability by 5 percent, the second by 5.25 percent, and the third by 5.26 percent. In this case the gain in reliability rapidly decreases with an increase in the degree of redundancy. Qualitatively, it is correct for any other situation, subject to the validity of the assumptions made above. Clearly, the second degree of redundancy (three units in parallel) is justified only when safety is of paramount importance.

Consider a single power generating unit again and assume that there are no concerns with respect to the reliability of the diesel. Then the unit can be designed in various other configurations. In Fig. 2-7 three possible configurations and corresponding reliability diagrams are shown. In the configuration of Fig. 2-7b it is assumed that the generator is the least reliable element and a redundant generator is used, whereas

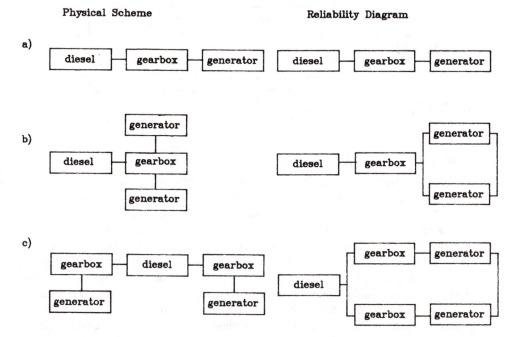

Figure 2-7 Physical arrangements and reliability diagrams of three diesel-generator systems.

in Fig. 2-7c a combination gearbox-generator is used as a redundant element. These are various levels of redundancy: from one redundant element to two redundant elements. The question is, which is better—to have redundant components or redundant subsystems? In other words, how is reliability affected by the level of redundancy?

For the three configurations shown in Fig. 2-7, if it is assumed that the gearboxes and generators have the same reliability functions irrespective of their parameters and also that failures of all components are independent, then a comparison of their reliability characteristics can be made using Eqs. (2-6) and (2-10). Assume that at the end of 1 year of operation the reliabilities of the diesel, gearbox, and generator are $R_d = 0.999$, $R_{gb} = 0.98$, and $R_{gr} = 0.97$, respectively. The reliabilities of the three configurations are then $R_a = 0.950$, $R_b = 0.978$, and $R_c = 0.996$. The conclusion can be made that the different levels of redundancy have different effects on the configuration's reliability. The degree of effect, however, should be assessed in any specific situation. Also, in this case, it was found to be beneficial to have redundant gearbox-generator units rather than generators only. However, it cannot be generalized that a high level of redundancy is always better. Actually, the opposite is true. If, instead of the reliability diagram Fig. 2-7b, an alternative (Fig. 2-8) was mechanically possible (or meaningful), then $R_d = 0.9977$, which is higher than for any other configuration. In Fig. 2-8 there is a true redundancy of components as opposed to the redundancy of units in Fig. 2-7c. However, in mechanical engineering, redundancy of components in most situations is difficult to implement. Besides, the gain may be offset by common-cause failures. For these reasons the redundancy of units is used more often than the redundancy of components.

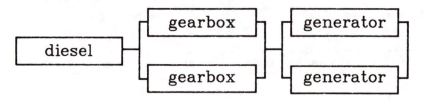

Figure 2-8 Reliability diagram of diesel-generator system with redundant components.

Markov Process of Random Events

In the analysis of systems based on reliability block diagrams it was assumed that the component failures are statistically independent and that they have a common time base. There are situations when these assumptions are not valid in principle. Then, even a comparison of various configurations cannot be done properly. A system with load-sharing components (or units) is an example of when a failure of one parallel component affects the level of "load" and thus the rate of failure of others. A system with a standby component (or a unit) is another example of when a time base is not the same, since a standby component begins operating only when the regular one fails.

The Markov technique expands the reliability analysis to statistical- and time-dependent systems. The limitation of the method is that it assumes a constant failure rate for any component. However, it will be shown in Chapter 4 that the constant failure rate is the most important from a practical standpoint. For a rigorous introduction to the Markov method, the reader is referred to special publications (see Appendix A). Herein the presentation is limited to the system of two components with constant but dependent failure rates. The objective of the following discussion is twofold: to illustrate the method and to indicate the effects of statistical dependency and a variable time base on system reliability.

Consider a two-component system, a and b being components, without specifying the components' reliability block diagram. The task is to evaluate the reliability of the system at time t_1. Up to this time there are four states possible for this system (Fig. 2-9). In state 1 two components operate successfully over the entire time span t_1. In state 2 two components operate only part of this time, since at some random point in time, component b has failed, so that only a is left. State 3 is similar to state 2 with the exception that b is left. State 4 constitutes a failure of the system at time t_1. Note, however, that states 2 and 3 may also constitute a system failure if two components are needed for the system to survive. Let us assume that only state 4 constitutes the system failure. Then the other states constitute three possibilities of successful operation during time t_1. Since these three states are mutually exclusive, the probability of success (the reliability) is equal to the sum of probabilities of being in one of the three states:

$$R(t_1) = P_1 (T > t_1) + P_2 (T > t_1) + P_3 (T > t_1) \tag{2-17}$$

where T is the time to failure.

The transition from one state to another is taking place at random moments in time. One of the basic assumptions of the Markov method is that the probability of transition from one state to the next is proportional to the time increment Δt over which the next event may occur. Thus we can relate the probability increment to the time increment.

Consider first the probability of staying in state 1. Suppose that there were no failures up to time $t < t_1$ and that the probability of this event is $P_1(t)$. If no failures took place during time increment Δt, then at the end of this time increment the probability of no failures would be $P_1(t + \Delta t)$. In order for this situation to be true, two events should take place simultaneously, namely, no failures by time t and no failures during time Δt. These two events are independent, so the probability of their occurrence is found by multiplying the probabilities that each of them will occur. Let us identify these two probabilities. The probability of no-failure by time t is $P_1(t)$. The probability of failure during time Δt is, in turn, determined by two possible events (see Fig. 2-9): the probability of failure of component a, which is equal to $h_a \Delta t$, and component b, which is equal to $h_b \Delta t$ (where h_a and h_b are proportionality constants characterizing the failure rates). These two events are also independent, and thus the probability of no-failure during time Δt is equal to $(1 - h_a \Delta t)(1 - h_b \Delta t)$. From the foregoing considerations, it must be clear that the following equality of probabilities holds:

$$P_1(t + \Delta t) = (1 - h_a \Delta t)(1 - h_b \Delta t)P_1(t) \qquad (2\text{-}18)$$

By multiplying the expressions in parentheses and neglecting the second-order term for Δt, the following finite difference equation is obtained:

$$\frac{P_1(t + \Delta t) - P_1(t)}{\Delta t} = -(h_a + h_b)P_1(t) \qquad (2\text{-}19)$$

which, when $\Delta t \to 0$, is transformed into the differential equation,

$$\frac{dP_1}{dt} = -(h_a + h_b)P_1 \qquad (2\text{-}20)$$

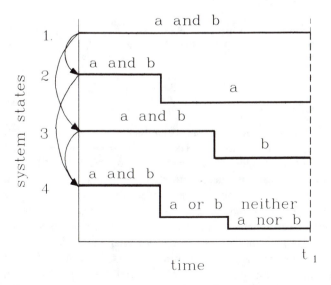

Figure 2-9 Probable states for a two-component system.

Equation (2-20) can be easily integrated, subject to the following initial conditions: when $t = 0$, $P_1 = 1$. As a result, the following expression for $P_1(t)$ is obtained:

$$P_1(t) = \exp[-(h_a + h_b)t] \tag{2-21}$$

So $P_1(t)$ is the probability of staying in state 1.

The probability of transition from state 1 to state 2 is determined by the probabilities of two mutually exclusive events: either the transition takes place before time t or during time Δt. The probability of the first event is a product of two probabilities: (1) failure before time t, $P_2(t)$, and no-failure of component a during time Δt, $(1 - h_a^* \Delta t)$, and (2) no-failure at time t, $P_1(t)$, and failure of the component b during time Δt, $h_b \Delta t$.

So the probability of transition to state 2 is

$$P_2(t + \Delta t) = (1 - h_a^* \Delta t)P_2(t) + h_b \Delta t P_1(t)$$

where h_a^* is the increased rate of failure of component a given that component b has failed.

The corresponding differential relationship is as follows

$$\frac{dP_2}{dt} = -h_a^* P_2 + h_b P_1 \tag{2-22}$$

Equation (2-22) can be integrated subject to the initial condition: at $t = 0$, $P_2 = 0$ and taking into account Eq. (2-21) for $P_1(t)$. The following expression for $P_2(t)$ is obtained:

$$P_2(t) = \{\exp(-h_a^* t) - \exp[-(h_a + h_b)t)]\} \frac{h_b}{h_a + h_b - h_a^*} \tag{2-23}$$

Similarly, the probability of transition from state 1 to state 3 is

$$P_3(t) = \{\exp(h_b^* t) - \exp[-(h_a + h_b)t]\} \frac{h_a}{h_a + h_b - h_b^*} \tag{2-24}$$

where h_b^* is the increased rate of failure of component b given that component a has failed. The reliability of the system can now be determined according to Eq. (2-17), where $t_1 = t$.

An r-out-of-n Configuration

In a reliability diagram with parallel arrangement of components considered earlier, all components must fail before the system fails. There are, however, situations when the system fails even when some components in a parallel arrangement can still operate. For example, if r out of n batteries needed to start a locomotive are in good condition, then the locomotive can be started. If the number of discharged batteries is larger than $(n - r)$, then there is not enough power. The case considered before, when $r = 1$, becomes a particular case of this more general situation of redundancy. The configuration of the system remains the same no matter how many redundant components are needed for the system to survive. However, for the same system if fewer components satisfy the same requirements, say, in power supply, strength, or force, then the components must be different.

Intuitively it is felt that if $r = n$, i.e., the number of components to survive equals the maximum number of components in parallel, then the level of reliability is smaller than when $r < n$. At the other extreme, if $r = 1$, then the reliability should be the maximum possible. Note that the above is true only if the reliability functions of components are identical for the various design solutions.

Once again the reliability of the system depends not only on the reliability of components but also on the way these components are arranged into the system. Since the latter is decided at the conceptual stage, it is important to analyze advantages and disadvantages of configurations themselves in order to arrive at the best component-configuration combination. The Markov method described in the previous section for the two-component system can be used to illustrate the point.

Equation (2-17) gives the reliability of a two-component system when only one component is sufficient to keep the system operational. In the case when $h_a = h_b = h_a^* = h_b^* = h$, Eq. (2-17) is reduced to

$$R_1(t) = -e^{-2ht} + 2e^{-ht} \tag{2-25}$$

When two components are needed to keep the system operational, the reliability is represented by the function $P_1(t)$, Eq. (2-21), that is, when $h_a = h_b = h$,

$$R_0(t) = P_1(t) = e^{-2ht} \tag{2-26}$$

Since

$$\frac{R_1(t)}{R_0(t)} = 2e^{ht} - 1 \tag{2-27}$$

it is seen that the ratio in Eq. (2-27) is larger than unity at any time $t > 0$. It shows that indeed by allowing the system to operate with one failed component out of two originally installed, the reliability of the system is increased. Moreover, this increase does not remain the same but becomes more significant with time.

Another aspect of an r-out-of-n configuration is illustrated in the following example, in which, rather than the Markov process, a binomial distribution of events, which is a particular case of the Markov process, is used.

Example 2-3 A needed power for a new plane can be supplied either by four identical engines or by three also identical but different engines. If four engines are used, then only two would be needed for a plane to survive. In the case of three engines, only one would be needed to survive. It is known that the smaller power engines have reliability p after t hours, whereas more powerful new engines do not have statistical data but it is expected that their failure rate would be higher than that for a smaller engine. The question is, what engine is to be used?

To properly answer this question, all other aspects associated with two types of engines should be considered: costs, weight, size, efficiency, problems of maintenance, etc. However, let us assume that these aspects for both engines are comparable. The question is then reduced to the problem of reliability, but the solution remains uncertain due to the absence of statistical data about the second engine. Faced with this uncertainty, the original question can be reformulated, namely (1) which configuration is more reliable given that both types of engines have identical reliability properties, and (2) what is the lowest reliability allowed for the larger engine if the second configuration is more advantageous in other respects.

In order to answer these two new questions the expressions for reliabilities of two configurations should be derived first.

Let us consider the case of four engines numbered 1, 2, 3, and 4 and denote by E_1, E_2, E_3, and E_4 the events that corresponding engines succeed after t hours in service, and also denote by p the corresponding reliabilities of each engine at time t. Since any two out of four engines are sufficient to keep the plane flying, any one of six combinations: $C_{12} = E_1E_2$, $C_{13} = E_1E_3$, . . . , $C_{34} = E_3E_4$ satisfies the safety requirement. Then the reliability is the probability that any of these combinations takes place, namely,

$$R_s = P(C_{12} \cup C_{13} \cup \ldots \cup C_{34}) \tag{2-28}$$

In general, for identical and statistically independent components the probability of system success if only r out of n need to succeed is based on the binomial distribution and is given by the following summation of binomial probabilities:

$$R_s = P(r \leq j \leq n) = \sum_{j=r}^{n} \binom{n}{j} p^j (1 - p)^{n-j} \tag{2-29}$$

where

$$\binom{n}{j} = \frac{n!}{j! \, (n - j)!}$$

In the case of two engines out of four, Eq. (2-29) gives

$$R_{s4} = \binom{4}{2} p^2(1 - p)^2 + \binom{4}{3} p^3(1 - p) + \binom{4}{4} p^4 \tag{2-30}$$

whereas in the case of one engine out of three, Eq. (2-29) gives

$$R_{s3} = \binom{3}{1} p(1 - p)^2 + \binom{3}{2} p^2(1 - p) + \binom{3}{3} p^3 \tag{2-31}$$

Now, if the reliability of both types of engines p is the same after t hours in service, then assuming that $p = 0.9$, the reliability of the four-engine configuration is $R_{s4} = 0.9963$, whereas $R_{s3} = 0.999$. (Note that the 1-out-of-3 configuration is a conventional, completely redundant system whose reliability can be found using Eq. (2-10). Check that it gives the same result.) It is clear that a three-engine configuration is more reliable given that engines in two configurations are identical as far as their reliabilities are concerned. This answers the first question posed above.

A doubt has been expressed that the more powerful and recently developed engine has the same level of reliability as a less powerful one. If this is the case, then what would the lowest reliability be at which $R_{s4} = R_{s3}$? By solving the latter numerically, using Eqs. (2-30) and (2-31), it can be found that the reliability of the engine in a three-engine configuration must be not less than $p = 0.84$. It seems that it gives a significant margin to cover the uncertainty with respect to the more powerful engine.

Effect of Load Sharing on System Reliability

In the case of r-out-of-n components in a parallel arrangement it was assumed earlier that component failure does not affect the reliability functions of the remaining components. Very often this is not true. Even in the case of 2-out-of-4 engines considered in Example 2-3, a failure of one engine would result in increased load on the remaining engines. This increased load affects the reliability of the remaining engines, namely, it will decrease. In general, in a situation like this the more elements that fail, the larger the rate of reliability deterioration will be. So far, only the configuration itself plus the reliability of components determined the reliability function of the system. Dependence of reliability on the level of load becomes an additional factor that affects the system reliability function. Let us consider an example.

Example 2-4 A shaft is designed with two keys, while only one is sufficient to transfer the torque. However, if one of two fails, then the load on the other increases, and the failure rate increases as well. This situation is correctly simulated by the Markov process for a two-component system presented above. Since in this case it can be assumed that $h_a = h_b = h$ and $h_a^* = h_b^* = h^*$, where h^* is the increased rate of failure when the load is doubled, Eqs. (2-21), (2-23), and (2-24) are reduced to

$$P_1(t) = \exp(-2ht) \tag{2-32}$$

$$P_2(t) = [\exp(-h^*t) - \exp(-2ht)]\,\frac{h}{2h - h^*} \tag{2-33}$$

$$P_3(t) = [\exp(-h^*t) - \exp(-2ht)]\,\frac{h}{2h - h^*} \tag{2-34}$$

respectively, so that the reliability of the shaft with two keys according to Eq. (2-17) is

$$R(t) = -\frac{h^*}{2h - h^*}\exp(-2ht)\,\frac{2h}{2h - h^*}\exp(-h^*t) \tag{2-35}$$

As an example, let us consider the case when $h^* = 2h$, i.e., the failure rate is doubled when one key fails, which means that the failure rate is proportional to the load. In this case, Eq. (2-35) cannot be determined. This can be resolved by applying L'Hospital's rule with respect to h^* and then setting h^* to $2h$. The result is

$$R(t) = (1 + 2ht)\exp(-2ht) \tag{2-36}$$

If the failures of two keys were independent, then the reliability of the system would be [set $h^* = h$ in Eq. (2-35)]

$$R(t) = 2\exp(-ht) - \exp(-2ht) \tag{2-37}$$

which is the reliability of two statistically independent components operating functionally in parallel. Comparison of Eqs. (2-36) and (2-37) indicates that in a "long run" a shared-load system is less reliable.

System with Standby Components

A standby component is a special case of redundancy in which a component is activated only when an identical one has failed. A spare tire is an example of a standby component. The question of switching to a standby component is important in itself, since it may involve automatic, semiautomatic, or manual actions, each of which may fail. However, it is assumed here that the switching is perfect. From a functional point of view a standby component is in parallel to the system. However, for regular parallel arrangement of components in a block diagram the time base for components is the same (estimation of reliability function starts at the same moment in time), whereas for a standby component the time base is different and the moment in time when the standby component is activated is random. A standby component increases the reliability of the system. However, the method of assessing this increase and the degree of increase are different from those for a regular parallel arrangement of components.

The Markov process of random events for a simple case is used here to demonstrate the basic principles involved.

Example 2-5 Two compressors a and b are used to supply air for the air-supported structure. One compressor is in an operating mode, while the other is in a standby mode. If the operating compressor fails, the back-up compressor takes over. Let us assume that the switching process is 100% reliable. What is the reliability of the system of air supply if each compressor has a constant failure rate, h_a and h_b, respectively?

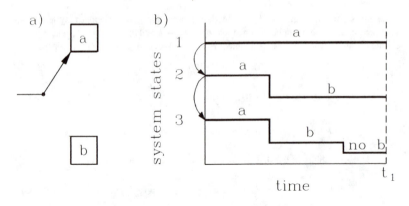

Figure 2-10 A two-component standby system and its probable states.

SOLUTION The schematic diagram of a standby system of two compressors, a and b, is shown in Fig. 2-10a. In Fig. 2-10b, two states define a successful system operation: (1) compressor a operates during time t_1, and (2) compressor a fails at $t < t_1$, but compressor b is switched in and operates successfully until $t = t_1$.

In order for the first state to be true, two events should take place simultaneously at the beginning of time interval t and at the end of time interval $t + \Delta t$: no-failure should take place before time t [the probability being $P_1(t)$], and no-failure should take place during time Δt [the probability being $(1 - h_a\Delta t)$]. Then the following equality of probabilities of no-failure at the end of the time interval $(t + \Delta t)$ holds:

$$P_1 (t + \Delta t) = (1 - h_a\Delta t)P_1(t) \tag{2-38}$$

where h_a is the constant failure rate of the compressor a. The corresponding differential relationship for Eq. (2-38) is

$$\frac{dP_1}{dt} = -h_a P_1 \tag{2-39}$$

Integrating Eq. (2-39) subject to initial conditions ($t = 0$, $P_1 = 1$) gives

$$P_1(t) = \exp(-h_a t) \tag{2-40}$$

The transition from state 1 to state 2 can take place either before time t [the probability being $(1 - h_b\Delta t)P_2(t)$] or during the time interval Δt [the probability being $h_a\Delta t P_1(t)$]. Then the following equality of probabilities holds:

$$P_2 (t + \Delta t) = (1 - h_b\Delta t)P_2(t) + h_a\Delta t P_1(t) \tag{2-41}$$

and the corresponding differential relationship is

$$\frac{dP_2}{dt} = -h_b P_2 + h_a P_1 \tag{2-42}$$

Integrating the latter subject to initial conditions ($t = 0$, $P_2 = 0$) and taking into account Eq. (2-40), the following is obtained:

$$P_2(t) = \frac{h_a}{h_b - h_a} [\exp(-h_a t) - \exp(-h_b t)] \tag{2-43}$$

Thus the reliability of the system with a standby compressor is

$$R(t) = P_1(t) + P_2(t) = \frac{1}{h_b - h_a} [(h_b \exp(-h_a t) - h_a \exp(-h_b t)] \qquad (2\text{-}44)$$

In our case two compressors are supposed to be identical, and it can be assumed that both have equal failure rates, $h_a = h_b = h$. In this case, Eq. (2-44) cannot be determined, since both numerator and denominator are equal to zero when $h_a = h_b$. L'Hospital's rule can be used to find $R(t)$ when $h_a = h_b$. In Eq. (2-44), h_b can be considered as a variable, which tends to h_a, while the latter is constant. By differentiating the numerator and denominator with respect to h_b and then letting $h_b = h_a = h$, the following expression is obtained:

$$R(t) = (1 + ht)e^{-ht} \qquad (2\text{-}45)$$

For $h = 1$ failure per year, the reliability functions with and without a standby compressor are shown in Fig. 2–11.

In summary, although reliability is not a major concern at the conceptual design stage, it should be used as one criterion for choosing the best alternative. Since the data base concerning information about components and their interaction is limited at this stage, a method of comparative assessment of various competing alternatives should be used. This comparison leads to an alternative with the increased potential for reliability. The problem of realizing this potential is left for the next two design stages.

2-4 PRELIMINARY DESIGN STAGE

The objective of the preliminary design stage is to show that a concept developed earlier can be materialized, which means that it can satisfy requirements for functional performance while meeting the constraints imposed on the product. Thus the emphasis shifts from meeting the functional requirements to meeting the constraint requirements

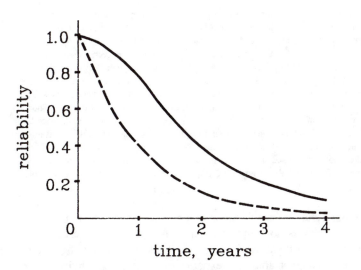

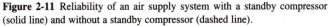

Figure 2-11 Reliability of an air supply system with a standby compressor (solid line) and without a standby compressor (dashed line).

such as size, weight, and cost. The functional requirement of a car is to meet specific payload capacity–speed requirements. This is done at the conceptual stage by choosing (or designing) an engine and transmission. A limit for the weight of the car is set as a target that should be met at later stages. The maximum speed (or acceleration at the start) is attained for a given engine-transmission system and load capacity if the weight of the car itself is as specified. Thus constraints cannot be set up arbitrarily; they originate from the requirement of product performance and if not adhered to would adversely affect performance. However, at the conceptual stage the weight of the car would be assumed based on past experience, similar products, weight of basic components, etc. (if any or all of this information is available). This assumed weight may come at the later stages, in conflict with other requirements such as size, shape, cost, fuel efficiency, and reliability. This is a typical design situation, and a compromise is often the only solution.

Thus product specifications set up as objectives of product design may be altered in the process of this design due to interdependence among various constraints. Reliability requirements, being part of these specifications, may undergo similar changes.

At the beginning of the preliminary stage, only input-output characteristics of the product are known. In the case of a car, the parameters of the engine (power, speed) would be known, while the dynamic parameters of the car (speed, acceleration) would be specified. The preliminary design is concerned with finding design solutions that materialize the input-output specification for a given configuration. The point is that the input-output specifications are given as nominal, not random, parameters, and the design engineer, while making decisions on type of parts, their geometry and location, and mutual arrangement, is using deterministic analyses of stresses, deflections, flows, speeds, etc., to come up with a product layout. At this initial moment of the second design stage, uncertainties, which are at the foundation of the reliability concept, are of no concern. Does it mean that the reliability is of no concern as well? The following example is intended to clarify this point.

Example 2-6 In Fig. 2-12 a spur gear–shaft system on two supports is shown. In Fig. 2-12a the gear is located in the middle of two supports, in Fig. 2-12b it is closer to the left support, and in Fig. 2-12c it is closer to the right support but cantilevered. Due to a radial force acting on the gear, the shaft will bend. In Fig. 2-12a the result of the bending will be radial displacement of the gear, whereas in Figs. 2-12b and 2-12c, displacement is accompanied by the gear rotation out of its plane. Both gear displacement and rotation affect the kinematics of teeth mesh; however, the gear rotation has a much more profound effect on redistribution of contact stresses and thus on the magnitude of these stresses. The magnitude of contact stresses, in turn, determines the lifetime, i.e., the reliability, of the gear. Thus a few design factors are important with respect to the reliability of a gear: shaft stiffness, distance between supports, and gear width and location. The design engineer knows that any solution that minimizes gear rotation and displacement increases gear reliability. So although the absolute level of gear reliability is not, and cannot be, assessed at this point due to lack of information, the relative level of reliability (or the potential for the reliability growth) can be increased by making proper design solutions.

Implementing the input-output specifications in a concrete design is by its nature an iterative process, since there is no unique sequence of steps that will result in the required solution. The number of iterations (or steps) depends on the task to be accomplished, state of knowledge in a specific area, experience of the design engineer, and available time and capital. Here the question of integrating the reliability aspects into this process is addressed.

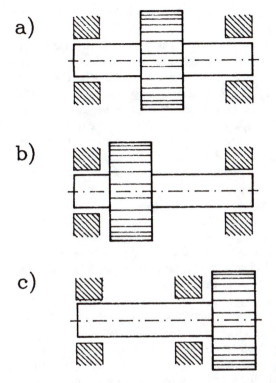

a)

b)

c)

Figure 2-12 Various design alternatives for spur gear–shaft assemblies.

At the conceptual stage the reliability potential of the configuration is the main concern. At the preliminary stage the emphasis is on the reliability of components. However, as mentioned above, the reliability of components is affected by the properties of the components themselves, the state of their interfaces with mating parts, the operational conditions of the system, the internal environment of the system, and also the external environment. So it is clear that the reliability of components cannot be found at this stage, since some of the information concerning components and most of the information concerning the system are not available. The objective again is to increase the relative level of reliability without assessing its absolute level. There are various means of achieving this goal such as better design solutions (see Example 2-6), minimization of the number of parts, and utilization of off-the-shelf parts of proven quality. However, the direct application of the principle of raising the relative level of reliability is found in the notion of the safety factor. Safety factor is a reflection of experience and knowledge accumulated in a specific area of engineering and as such it accommodates in a simple way the notion of reliability. However, it should be made clear that the safety factor is not a measure of reliability, although there is a correlation between the two (see next section). The safety factor is associated with the nominal parameters of the system at the beginning of the service, whereas reliability is concerned with the nominal parameters plus their scatter as functions of time in service. So even the degree of correlation between the safety factor and reliability does not remain the same in time, namely, it decreases with time. Nevertheless, at the preliminary design stage the concept of a safety factor can be used as a substitute of reliability because it increases the relative level of reliability, thus meeting the objective of design at this stage. The refinement of the design using proper reliability methods can be done later when more information becomes available.

In the process of arranging machine components in a system, two factors are important as far as reliability is concerned: (1) identification of critical components and design elements and (2) identification of uncertainties that may have the most profound effect on product operation. The first factor is associated with either new design solutions, new materials, or new requirements. This factor also carries an element of uncertainty and thus requires special attention in terms of either literature survey or analytical and numerical investigation. However, it does not necessarily involve a statistical approach but may only be the lack of information about, for example, properties of material at elevated temperatures and high pressure, or effect of geometry on a stress-strain state of a component. As long as this information becomes available, the conventional safety factor approach can be used.

The second factor is directly associated with component or interface reliability. A thorough analysis of all factors contributing to the scatter of design parameters requires understanding of basic engineering sciences combined with knowledge of the system being designed. The ability to foresee potential problems becomes crucial at this point. Usually an experienced design engineer performs an exercise equivalent to a mental simulation of product functioning, in essence, postulating various conditions ("what if" type questions) like change in temperature, load, effect of nonuniform shape, nonflatness, surface finish, and foreseeing their potential effect. The result of this exercise is the formulation of a need for a probablistic analysis, computer simulation, tests, or all of the above for identified components and interfaces. If such a need is satisfied, then at the next design stage it will allow proper formulation of specifications for product manufacturing. At the same time, if at the preliminary stage a proper analysis of potentially troublesome uncertainties is done, then at the next stage, only the refinement of specifications might be needed. Otherwise, either the potential flaw would slip unnoticed, or if discovered later, it would cause a delay in the design process due to its possible effect on other components.

Example 2-7 Consider a spline joint connecting two shafts (see Fig. 2-13). A design engineer might ask how the spline joint may fail and which of its failure modes are associated with material, geometry, interfaces, and operating conditions. A joint taken out of the assembly becomes an abstract entity in the sense that the influence of a system cannot be properly discussed. However, even in this isolated form, it reveals the presence of uncertainties important in predicting a spline joint functioning. There are a few failure modes of a spline joint: (1) the shaft itself may fail due to the stress concentration at the root of the spline, (2) the spline may break due to the bending stresses, (3) the spline may sheer off, and (4) the surface contact stresses may cause wear and surface fatigue. Which one of these four modes will prevail depends on uncertainties associated with material, geometry, interfaces, and operating conditions. However, it is clear that any failure mode depends on the conditions at the interface between the two mating splines. Ideally all splines in a joint are supposed to be taking part in a torque transfer. However, due to the axial and angular misalignment of two shafts and due to the tolerances on the spline width, the actual number of splines participating in a torque transfer may be very limited. This results in the overloading of a spline and nonuniform load distribution along the spline. Since spline couplings are often used to compensate for the misalignment, the design engineer might ask how much misalignment for the given operating conditions the joint might take without failure. This question should be addressed either to analytical, test, or reliability engineers, or to all of these. However, it is the design engineer who identifies the potential problem.

The question of foreseeing the sources of uncertainties is outside the scope of this book and is mentioned here for the sake of completeness. The effect of these uncer-

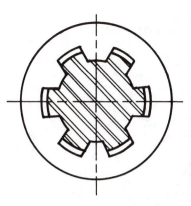

Figure 2-13 A spline coupling of two shafts.

tainties and how to take them into account is, however, of prime interest here and is considered next.

Distribution of Uncertainties and Reliability

The uncertainty in component performance follows from two "competing" factors: uncertain properties of the component itself and uncertainty in external loads. This is analogous to a relationship between stress and strength, where stress is associated with external loads and strength characterizes the property of the material.

Consider a spline joint again, as an illustration, and assume that a torque to be transmitted by this joint remains constant at a constant speed. Does it mean that, for example, the bending stress in a spline will remain the same in similar joints? As discussed in Example 2-7, due to random variation of the degree of misalignment, the distribution of forces acting on the splines will vary from one joint to another, and as a result, the maximum bending stress will vary.

Let us assume now that the load remains constant for all joints; however, the stress in each joint is different and randomly distributed with some density $f(\sigma)$. Then, if the ultimate strength of the material is constant for all similar joints, the distribution of stress-strength relationships for various joints is qualitatively shown in Fig. 2-14. The shaded area indicates those joints in which stress is larger than the strength and should be considered as failures. Thus the shaded area characterizes the probability of failure, and, as long as the stress probability density function and the magnitude of strength are known, the probability of failure (unreliability) can, in principle, be found.

Example 2-8 Consider again the spline joint and assume that the alignment of the joint is ideal. However, the spacing between the mating splines and grooves is random due to inaccuracies in manufacturing. Assume that the spline width is $b = b_{n-\Delta b}^{+0}$, and the width of groove is $a_{n-0}^{+\Delta a}$, where b_n and a_n are the nominal dimensions and Δb and Δa are the ranges of tolerances (see Fig. 2-15). In a perfectly aligned joint because of the scatter of spline thickness and width of groove dimensions, some mating interfaces may have clearances and thus be prevented from participating in a load transfer. The condition for the clearance is

$$c = a - b > 0 \tag{2-46}$$

If a and b are two random variables with corresponding distributions $f(a)$ and $f(b)$, then assuming that a distribution of c can be found, $f(c)$, the probability of having a clearance is

$$P(c > 0) = \int_0^\infty f(c)\,dc \tag{2-47}$$

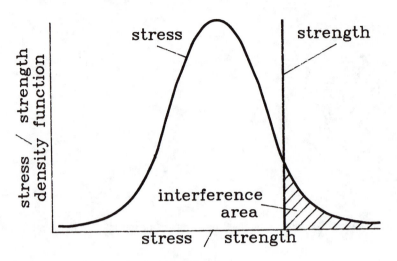

Figure 2-14 Illustration of stress-strength interference.

For a total number of splines in a joint N the probability that N_f of them will not be in contact is

$$P = \frac{N_f}{N} \tag{2-48}$$

Since probabilities in Eqs. (2-47) and (2-48) should be the same, P found from Eq. (2-47) on the basis of tolerances allows N_f to be determined in Eq. (2-48).

To make this example more meaningful, assume that Δa and Δb are normally distributed variables (which is a reasonable assumption for the distribution of tolerances in general) and the corresponding parameters (means and standard deviations) are as follows: for the width of the groove, μ_a and σ_a, and for the spline thickness, μ_b and σ_b. It can be taken that $\mu_a = a_n + \mu(\Delta a) = a_n + 0.5\Delta a$ and $\mu_b = b_n - \mu(\Delta b) = b_n - 0.5\Delta b$. It can also be assumed that half the tolerance range equals three standard deviations (the so-called "three-sigma" rule), namely, $3\sigma_a = 0.5\Delta a$ and $3\sigma_b = 0.5\Delta b$. Then a random variable c in Eq. (2-46) is also normally distributed with a mean

$$\mu_c = \mu_a - \mu_b = 0.5(\Delta a + \Delta b) \tag{2-49}$$

and a standard deviation

$$\sigma_c^2 = \sigma_a^2 + \sigma_b^2 = \frac{0.25}{9}(\Delta a^2 + \Delta b^2) \tag{2-50}$$

In Eq. (2-49) it is taken that $a_n = b_n$.

The distribution function for c is

$$f(c) = \frac{1}{\sigma_c\sqrt{2\pi}} \exp\left[-\frac{1}{2}\left(\frac{c - \mu_c}{\sigma_c}\right)^2\right] \qquad -\infty < c < \infty \tag{2-51}$$

This function is shown in Fig. 2-16. The probability of clearance (shaded area in Fig. 2-16) is found from Eq. (2-47):

$$P(c > 0) = \int_0^\infty \frac{1}{\sigma_c\sqrt{2\pi}} \exp\left[-\frac{1}{2}\left(\frac{c - \mu_c}{\sigma_c}\right)^2\right] dc \tag{2-52}$$

Performing the substitution of variables, namely, letting $z = (c - \mu_c)/\sigma_c$, the integral in Eq. (2-52) is transformed to

$$P(z_0) = \frac{1}{\sqrt{2\pi}} \int_{z_0}^\infty \exp\left(-\frac{z^2}{2}\right) dz \tag{2-53}$$

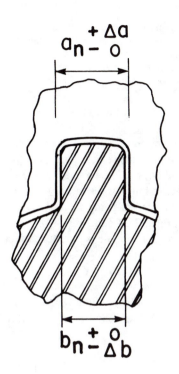

Figure 2-15 Spline coupling geometry.

where

$$z_0 = -\frac{\mu_c}{\sigma_c} = -\frac{\mu_a - \mu_b}{\sqrt{\sigma_a^2 + \sigma_b^2}} = -3\frac{\Delta a + \Delta b}{\sqrt{\Delta a^2 + \Delta b^2}} \tag{2-54}$$

The integral in Eq. (2–53) can be found using normal tables (see Appendix A). It is seen that its value depends on the parameter z_0. Let us consider a particular case when $\Delta a = \Delta b$. Then $z_0 = -3\sqrt{2} = -4.24$ in Eq. (2-54). The probability of clearance is $P(c > 0) = \phi(4.24)$ and is close to 1, which means that the probability of having a sliding fit is close to zero, which in turn, means that only one

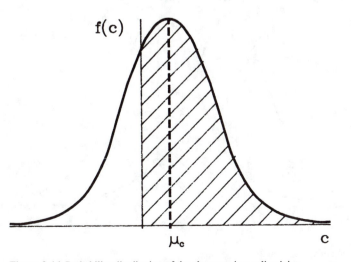

Figure 2-16 Probability distribution of the clearance in a spline joint.

spline will be engaged. This is understandable, since for a continuous distribution of clearances the probability of having two equal minimum clearances (and correspondingly, two engaged splines) is zero. However, the true number of splines participating in a torque transfer may be larger than 1 if the flexibility of splines is taken into account and if the latter is sufficiently large. The number of splines involved in a torque transfer, and thus the stress level, depends on geometry and external load.

Example 2-8 illustrates that even if the external load is constant, the stress caused by this load may be essentially random.

Referring again to Fig. 2-14, let us drop the assumption of constant strength and assume, instead, that strength is randomly distributed with some probability density function (pdf) $f(s)$. The combined stress-strength distribution is shown in Fig. 2-17. It is seen that since there is a probability of strength being lower (broken line) than its constant mean value (solid line), the probability of the stress being larger than the strength has been increased (compare shaded areas in Figs. 2-14 and 2-17).

To determine formally the probability of failure, a new variable is introduced in a form

$$y = s - \sigma \tag{2-55}$$

where s refers to strength and σ refers to stress. Since s and σ are both random variables, y is also a random variable described by the pdf $f(y)$. The condition of no-failure is $y > 0$, and the probability that it will occur is

$$P(y > 0) = \int_0^\infty f(y)\, dy \tag{2-56}$$

Usually the information related to the distribution functions for a particular design is not known beforehand. Even if similar design cases are well documented, there are always particular circumstances affecting the distribution functions. For example, even if shaft designs are identical, the steel may be supplied by different manufacturers, the shafts may operate under different load and environmental conditions, or they may be

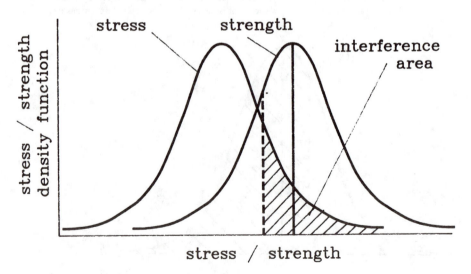

Figure 2-17 Illustration of stress-strength interference.

manufactured at different plants. As a result, the stress and strength distribution functions may be affected. The simplest and most common case is when stress and strength are normally distributed. (Note that stress and strength should be understood here in a general sense; for example, "stress" can be an elongation of a turbine blade due to creep at elevated temperatures, and "strength" is then the clearance between the stator and rotor, or it can be a corrosion-environment interference, etc.)

Normally distributed stress and strength are written as

$$f(\sigma) = \frac{1}{\sigma_\sigma \sqrt{2\pi}} \exp\left[-\frac{1}{2}\left(\frac{\sigma - \mu_\sigma}{\sigma_\sigma}\right)^2\right] \qquad -\infty < \sigma < \infty \qquad (2\text{-}57)$$

and

$$f(s) = \frac{1}{\sigma_s \sqrt{2\pi}} \exp\left[-\frac{1}{2}\left(\frac{s - \mu_s}{\sigma_s}\right)^2\right] \qquad -\infty < s < \infty \qquad (2\text{-}58)$$

respectively, where μ_σ and μ_s are the mean values of the stress and strength and σ_σ and σ_s are the standard deviation of the stress and strength.

The variable y in Eq. (2-55) is also normally distributed with a mean

$$\mu_y = \mu_s - \mu_\sigma \qquad (2\text{-}59)$$

and a standard deviation

$$\sigma_y^2 = \sigma_s^2 + \sigma_\sigma^2 \qquad (2\text{-}60)$$

The distribution function then is

$$f(y) = \frac{1}{\sigma_y \sqrt{2\pi}} \exp\left[-\frac{1}{2}\left(\frac{y - \mu_y}{\sigma_y}\right)^2\right] \qquad -\infty < y < \infty \qquad (2\text{-}61)$$

This function is shown in Fig. 2-18. The reliability, after the substitution of Eq. (2-61) into Eq. (2-56) and transformation of variables similar to that in Eq. (2-52), is given by

$$R(z_0) = \frac{1}{\sqrt{2\pi}} \int_{z_0}^{\infty} \exp\left(-\frac{z^2}{2}\right) dz \qquad (2\text{-}62)$$

where

$$z_0 = \frac{-(\mu_s - \mu_\sigma)}{\sqrt{\sigma_s^2 + \sigma_\sigma^2}} \qquad (2\text{-}63)$$

It is seen that the reliability is expressed in terms of a parameter z_0 and can be found in the form

$$R(z_0) = 1 - \phi(z_0) \qquad (2\text{-}64)$$

where

$$\phi(z_0) = \frac{1}{\sqrt{2\pi}} \int_{-\infty}^{z_0} \exp\left(-\frac{z^2}{2}\right) dz \qquad (2\text{-}65)$$

is tabulated (see Appendix A).

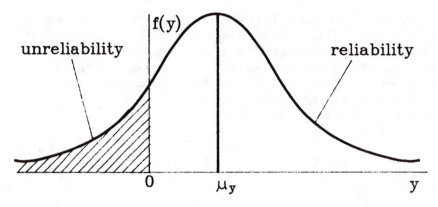

Figure 2-18 Probability density function.

Equation (2-63) is called the coupling equation, since it interrelates reliability with stress and strength. Note that Example 2-8 gives an application of this interrelationship in which "stress" is represented by the width of the groove and "strength" is represented by the spline thickness. The following example further demonstrates the application of the stress-strength interference approach in reliability assessment.

Example 2-9 In Example 1-4 the torque transmitted by a force fit is equal to

$$T = A\Delta \tag{2-66}$$

where

$$A = 0.25f\pi DLE(1 - D^2/4b^2) \tag{2-67}$$

Δ is the interference of tolerances, and the other notation can be found in Example 1-4. Assume that A is constant. Let us represent the shaft and hole radii as follows:

$$R_s = R_{sn} + \Delta s \tag{2-68}$$

$$R_h = R_{hn} - \Delta h \tag{2-69}$$

where R_{sn} and R_{hn} are the nominal radii for the shaft and hole and Δs and Δh are the small deviations within the limits of tolerances (Δs and Δh are random variables). The interference Δ is equal to

$$\Delta = R_s - R_h = (R_{sn} - R_{hn}) + \Delta s + \Delta h = \Delta_{min} + \Delta s + \Delta h \tag{2-70}$$

where Δ_{min} is the minimum interference. Assume that Δs and Δh are normally distributed with the means and standard deviations as follows: $\mu_{\Delta s}$, $\mu_{\Delta h}$, $\sigma_{\Delta s}$, $\sigma_{\Delta h}$. It follows that Δ is also a random variable distributed with the following mean and standard deviation:

$$\mu_\Delta = \Delta_{min} + \mu_{\Delta s} + \mu_{\Delta h} \tag{2-71}$$

$$\sigma_\Delta^2 = \sigma_{\Delta s}^2 + \sigma_{\Delta h}^2 \tag{2-72}$$

The torque, Eq. (2-66), is also a random variable normally distributed with

$$\mu_T = A\mu_\Delta \tag{2-73}$$

$$\sigma_T = A\sigma_\Delta \tag{2-74}$$

The design is reliable if the external torque T_e is smaller than a torque carrying ability of the force fit, i.e., $T_e < T$. Let us assume that T_e is also normally distributed with μ_{T_e} as a mean and σ_{T_e} as a standard deviation.

For the given conditions it is of interest to find a relationship between the reliability of a force fit and a varying value of the length of the gear disk–shaft interface L (see Fig. 1-2; let us assume there is no freedom in varying other parameters of the force fit). The following data are given:

$$f = 0.1 \qquad E = 0.2 \times 10^4 \text{ MN/m}^2$$
$$D = 0.06 \text{ m} \qquad b = 0.12 \text{ m}$$
$$\Delta_{min} = 0.01 \times 10^{-3} \text{ m}$$
$$\mu_{\Delta s} = 0.015 \times 10^{-3} \text{ m} \qquad \mu_{\Delta h} = 0.020 \times 10^{-3} \text{ m}$$
$$\sigma_{\Delta s} = 0.010 \times 10^{-3} \text{ m} \qquad \sigma_{\Delta h} = 0.014 \times 10^{-3} \text{ m}$$
$$\mu_{Te} = 10^{-5} \text{ MN m} \qquad \sigma_{Te} = 5 \times 10^{-6} \text{ MN m}$$

SOLUTION The reliability is determined by the coupling equation

$$z_0 = \frac{-(\mu_T - \mu_{Te})}{(\sigma_T^2 + \sigma_{Te}^2)^{1/2}}$$

Substitution of parameters μ_T and σ_T, using Eqs. (2-73) and (2-74) and the corresponding numerical data, gives

$$z_0 = \frac{-(400L - 10)}{[(150L)^2 + 25]^{1/2}}$$

The reliability of the force fit when L varies from $L = R$ to $L = 4R$ ($R = 0.5D$) is shown in Fig 2-19. It is seen that when the length/radius ratio $L > 2$ is exceeded, the increase in reliability becomes marginal. A parametric study such as this one may be useful in many design situations. In conducting such analyses, the validity of assumptions for different combinations of system parameters should be questioned. Note in this respect that when $L > R$, the ratio of the actual area of contact to the nominal area between the shaft and the hub becomes smaller, which means that the coefficient of friction becomes smaller and the analysis of reliability for $L/R > 1$ should take this into account.

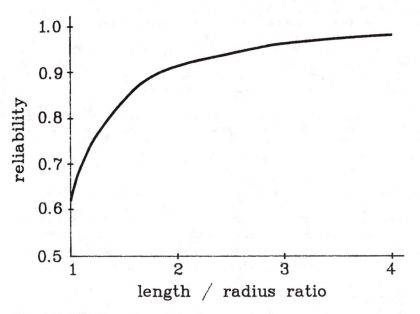

Figure 2-19 Reliability versus press fit length.

Safety Factors and Reliability

Conventionally, the safety factor is defined as the ratio of nominal strength to nominal stress. Since the nominal values in most cases coincide with the mean values, the safety factor can be written as

$$n = \mu_s/\mu_\sigma \qquad (2\text{-}75)$$

Using notation Eq. (2-75), the coupling equation Eq. (2-63) and thus the reliability can be expressed in terms of the safety factor and factors of variability of stress and strength in the following way:

$$z_0 = \frac{-(n-1)}{\sqrt{(n^2 V_s^2 + V_\sigma^2)}} \qquad n \geq 1 \qquad (2\text{-}76)$$

where $V_s = \sigma_s/\mu_s$ is the factor of variability of strength and $V_\sigma = \sigma_\sigma/\mu_\sigma$ is the factor of variability of stress.

Equation (2-76) shows that for a constant safety factor n the reliability of the system changes due to the change of variability factors, namely, when variability increases, the reliability of the system decreases. It means that the conventional way of assessing the design is not adequate with respect to the statistical scatter of the design parameters.

Besides allowing a more accurate assessment, provided the corresponding distribution functions are known, probabilistic design approach brings to the attention of a designer the parameters most influencing the reliability of the system. As a result, justified tolerances and specifications can be worked out. At the later stages, cooperation between the designer and the reliability engineer will be needed to finalize the drawing requirements by using a more thorough probabilistic analysis.

> **Example 2-10** The press fit analyzed before is considered again. The objective is to show how the variability of the interference affects the reliability while the safety factor remains the same.
>
> In this example the length of the interface L is considered to be fixed and $L = D = 0.06$ m. For the data given in Example 2-9 the safety factor of the press fit is $n = \mu_T/\mu_{Te} = 2.5$, and the corresponding reliability is $R = 0.908$. Now let us assume that wider tolerances on the shaft and the hole were assigned and, as a result, standard deviations had changed to the following: $\sigma_{\Delta s} = 0.012 \times 10^{-3}$ m and $\sigma_{\Delta h} = 0.016 \times 10^{-3}$ m. What is the effect of the new tolerances on reliability?
>
> SOLUTION Using the formula in Example 2-9, $\sigma_T = 0.0106 \times 10^{-3}$ m. Other parameters being the same, the reliability is found from the coupling equation to be $R = 0.88$. It is seen that for the same safety factor an increase in variability results in the decrease of reliability by 3 percent. Note that the degree of variability depends on the margin of tolerances.

Stress-Strength Interference Concept

In the expression for reliability given by Eq. (2-62) it was assumed that both load and resistance are time independent. As was discussed in Chapter 1, this combination of load and resistance does not belong to reliability engineering in a strict sense because reliability is defined as probability of success in time. However, the concept of load-resistance interference can be extended to include reliability as a changing function of time for both instantaneous and gradual failures.

Consider the case of an instantaneous failure of a spline joint when stress and strength are described according to Fig. 2-14. Note, however, that in Fig. 2-14 the torque causing the stress was assumed to be constant and the shaded area indicated the probability of failure in time zero. Let us assume now that the torque is a random function of time (see Fig. 2-20). Then for two different torques T_1 and T_2 the interference of stress and strength can be seen from Fig. 2-21. Each level of torque is uniquely related to the probability of success (reliability). The time when a specific level of torque is reached is a random variable (see Fig. 2-20). The probability of failure equals the product of probabilities of two events:

$$F_s(t) = P(T > T_j) \, P(\sigma > s) \tag{2-77}$$

where $P(T > T_j)$ is the probability of a torque exceeding the given torque T_j and is a function of time, and $P(\sigma > s)$ is the probability of stress exceeding strength. To determine the former, the theory of random functions should be used (although the solutions for random stationary processes are known, they are too complicated to be included in this introductory text). What is important here is that if the external torque is not constant in time, then the stress-strength interference becomes a function of time and thus there is the probability of failure. Thus the time dependence of external forces becomes a factor that transforms a problem of probabilistic design into one of reliability.

In general, both load and resistance are functions of time more often than not. The aging that takes place in any system due to corrosion, wear, fatigue, diffusion, etc., constitutes the decaying resistance. For example, in the case of a ball bearing the balls become plastically deformed in time, the races will have pits and scourings, and cracks may develop on rings and balls. The rate of ball bearing deterioration depends on the properties of the bearing itself but also on the properties of the system (lubrication, temperature, vibrations, alignment) and the fluctuation of the external load. Thus the rate of deterioration may vary from one product to another for the same type of bearing. However, if the distribution of these rates is known, then in principle, the load-resistance interference theory can be applied to determine the reliability of the

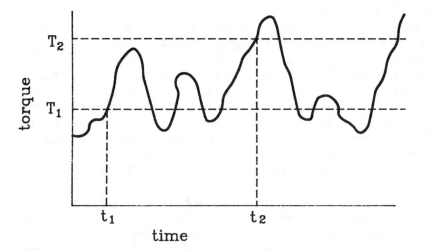

Figure 2-20 Time dependence of torque.

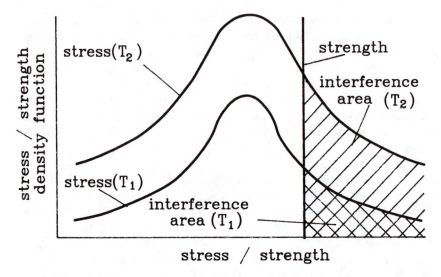

Figure 2-21 Illustration of stress-strength interference for various external torques.

product as a function of time. The parenthetical clause "in principle" is used because the physical laws of deterioration are very complicated, and so the distributions of properties are usually not available at the time of design. The design engineer tends at this stage to increase the relative level of reliability by identifying the causes of deterioration and failure and by trying to minimize their effect.

2-5 DETAILED DESIGN STAGE

The objective of the detailed design stage is to give detailed instructions on how to make a product. These instructions in the form of drawings and specifications indicate requirements to the geometry, materials, heat treatment, surface finish, etc. Thus it seems that the problem of achieving the functional objectives of the product and meeting the imposed constraints is over before the detailed design starts. This is not so. For example, the nominal efficiency of the internal combustion engine is determined at the conceptual stage of engine design. However, some of the factors affecting the actual efficiency are the friction losses and the sealing between piston and cylinder. These losses and the effectiveness of the sealing depend on the piston-cylinder interface and specifically on the quality and geometry of the spring rings. Although the nominal value of these losses can be estimated at the preliminary stage, the scatter of these losses can be estimated only when the geometrical tolerances for the piston, rings, and cylinder are assigned at the detailed design stage. Thus the refinement of engine efficiency is also done at the detailed design stage.

Tolerances assigned to components are subject to two requirements: to keep the functional performance of the product within acceptable limits and to meet the manufacturing constraints, i.e., the machine tools should be able to produce a part according to specifications. These two requirements are always in conflict, and a compromise is achieved on the basis of cost considerations. The first requirement includes reliability. More specifically, since the scatter of product failures is a reflection of the scatter of product properties, and the latter is, to a great extent, a result of random variations within assigned tolerances, the act of assigning tolerances has a profound effect on reliability and in fact finalizes reliability. Thus tolerances cannot be assigned arbitrarily even if they are within the capability of the machine tools. Of course, the effect of the scatter of properties of materials and the effect of the manufacturing processes on these properties are also important when evaluating the reliability of the product.

So by specifying geometry, materials, and manufacturing processes, a product as a system of interacting components is finalized. If at the conceptual design stage the design engineer was concerned with the relative level of reliability of a configuration, and at the preliminary stage with the relative level of reliability of components, at the detailed design stage the design engineer in principle can estimate the absolute level of reliability of a product as a system of interacting components. Thus the reliability integration process comes back to evaluation of the reliability of a system, however, on a different level of information about components and interfaces.

Example 2-11 The rotational component of the dynamic load on the supports of a hydroelectric unit is caused by the corresponding hydrodynamic, electromagnetic, and mechanical components. The hydrodynamic rotational force is caused by the eccentricity of the hydraulic turbine axis with respect to its axis of rotation (see Fig. 2-22). Similarly, the rotational electromagnetic force is caused by the eccentricity of the generator axis with respect to its axis of rotation. In both cases there is a nonuniformity of the clearance between the rotor and the stator, the result of which is the rotation of the minimum clearance. The forces of the mechanical unbalance are caused by the displacement of the center of mass with respect to the axis of rotation. Thus, if it was not for manufacturing or assembly errors there would be no rotational dynamic loads on supports and, as a result, no structural vibrations. The problem of establishing a relationship (in a statistical sense) between the tolerances and resulting

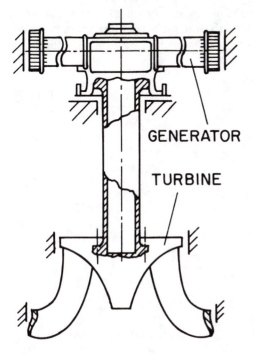

GENERATOR

TURBINE

Figure 2-22 Schematic diagram of hydraulic turbine-generator unit.

unbalanced forces is very important, since it allows a design engineer to properly specify tolerances. Note that unbalanced dynamic forces affect the reliability of bearings and other components susceptible to vibrations. It is seen that the above problem of estimating the unbalanced forces cannot be solved by considering only one part or even one element of the system (like a turbine); rather, the entire system, including its rotational and stationary components, should be analyzed. It could not be done at the conceptual or at the preliminary design stages because the component tolerances were not known at that time.

Random Nature of Unbalanced Forces

Unbalanced forces are present in all rotors, and so their cause and random nature should be clearly understood by a design engineer.

In Fig. 2-23 the shaded area indicates (in an exaggerated form) the true cross section of the shaft, whereas the dashed line circles indicate the limits within which the shaft diameter may actually be. Since the cross section of the shaft is never a perfect circle, the center of its mass will be displaced with respect to the center of rotation on a magnitude e. The value of e is random, since e is directly related to the tolerance Δ (see Fig. 2-23). Namely, if the shaft is rotated and a gauge is placed radially in a given cross section to its center of rotation measuring the surface displacement, then the range of variation of surface displacement should not exceed $\Delta/2$. At the same time this range is equal to $2e$. This leads to a relationship between the tolerance and the mean of the eccentricity magnitude

$$\mu_e = \Delta/4 \tag{2-78}$$

The same arguments are valid for a hollow shaft as well. In this respect, let us consider a general problem of n assembled rings (in Fig. 2-24, only two rings are shown). Assume that the external surface of each ring is displaced with some eccentricity with respect to the internal surface, whereas the mating surfaces of all rings are co-axial. Then the eccentricity of the nth ring with respect to the axis of rotation (assume that the latter coincides with the geometrical axis of the smallest inner ring) is equal to the sum of n randomly oriented eccentricity vectors,

$$\bar{e} = \sum_{i=1}^{n} \bar{e}_i \tag{2-79}$$

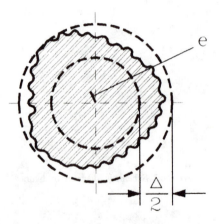

Figure 2-23 Illustration of nonroundness within the tolerances.

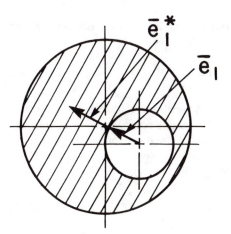

Figure 2-24 On the eccentricity of nonideal rings.

The displacement of the center of mass of the first ring $\bar{e}_1^*$ is linearly dependent on the eccentricity vector for this ring:

$$\bar{e}_1^* = \gamma_1 \bar{e}_1 \tag{2-80}$$

The displacement of the center of mass of the second ring is equal to the eccentricity of the first ring plus its own displacement of the center of mass:

$$\bar{e}_2^* = \bar{e}_1 + \gamma_2 \bar{e}_2 \tag{2-81}$$

By induction for the kth ring, its own displacement of the center of mass with respect to the center of rotation will be

$$\bar{e}_k^* = \sum_{i=1}^{k-1} \bar{e}_1 + \gamma_k \bar{e}_k \tag{2-82}$$

The radius of the center of mass of the entire assembly (the resultant eccentricity vector) is then

$$\bar{\rho}_n = \left(\sum_{k=1}^{n} m_k \bar{e}_k^* \right) \left(\sum_{k=1}^{n} m_k \right)^{-1} \tag{2-83}$$

where m_k is the mass of the kth ring.

In Eq. (2-83), $\bar{\rho}_n$ is the sum of vectors with random magnitudes and orientations. If all $\bar{e}_k^*$ are substituted, then Eq. (2-83) can be written as a linear combination of geometrical eccentricities with known constant coefficients as follows:

$$\bar{\rho}_n = \sum_{i=1}^{n} \alpha_i \bar{e}_i \tag{2-84}$$

where

$$\alpha_i = (m_i \gamma_i + m_{i+1} + \cdots + m_n) \left(\sum_{k=1}^{n} m_k \right)^{-1} \tag{2-85}$$

What is of interest in this problem is to find the pdf of the absolute value of $|\bar{\rho}_n|$ given the distribution function of vectors $\bar{e}_i$. For a particular case of the distribution of $\bar{e}_i$, this can be done explicitly.

The pdf of the amplitude of the eccentricity vector $e_i = |\bar{e}_i|$ is described by the Rayleigh distribution (see Fig. 2-25):

$$f(e) = \frac{e}{\sigma_i^2} \exp\left(-\frac{e^2}{2\sigma_i^2}\right) \tag{2-86}$$

where σ_i is the parameter of distribution and e is a variable eccentricity. It is important to note that if e_i is Rayleigh distributed, then the x and y components of the vector $\bar{e}_i$ are normally distributed with the mean $\mu_{ex} = \mu_{ey} = 0$ and the variance $\sigma_{ex}^2 = \sigma_{ey}^2 = \sigma_i^2$.

Consider then the x and y components of $\bar{\rho}_n$ in Eq. (2-84):

$$\rho_{nx} = \sum_{i=1}^{n} \alpha_i e_{ix} \tag{2-87}$$

and

$$\rho_{ny} = \sum_{i=1}^{n} \alpha_i e_{iy} \tag{2-88}$$

Since e_{ix} (correspondingly e_{iy}) are independent random variables that can be considered to be normally distributed, then ρ_{nx} (correspondingly ρ_{ny}) is also normally distributed with the following parameters:

$$\mu_{\rho x} = \mu_{\rho y} = 0 \qquad \sigma_{\rho x}^2 = \sigma_{\rho y}^2 = \sum_{i=1}^{n} \alpha_i^2 \sigma_i^2 \tag{2-89}$$

These relationships mean that there is some Rayleigh distribution for the resultant eccentricity vector such that its x and y components are normally distributed with the parameters according to Eq. (2-89). It follows that the Rayleigh distribution of the amplitude of the resultant eccentricity vector $\rho_n = |\bar{\rho}_n|$ is

$$f(\rho_n) = \frac{1}{\sigma^2} \rho_n \exp\left(-\frac{\rho_n^2}{2\sigma^2}\right) \tag{2-90}$$

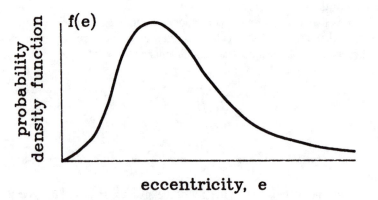

Figure 2-25 Rayleigh probability density function.

where

$$\sigma^2 = \sum_{i=1}^{n} \alpha_i^2 \sigma_i^2 \tag{2-91}$$

Considering that a mean of the Rayleigh distribution is equal to $(\pi/2)^{1/2}\sigma_i$ for each ring and that it should be equal to μ_e according to Eq. (2-78) a relationship between the tolerance for each outside diameter of a ring Δ_i and the distribution parameter σ_i is found

$$\sigma_i = \left(\frac{2}{\pi}\right)^{1/2} \frac{\Delta_i}{4} \tag{2-92}$$

Thus the distribution of the center of mass of the stack of rings is fully defined in terms of tolerances assigned to individual rings.

For a constantly rotating stack of rings the centrifugal force is

$$P_c = m\omega^2 \rho_n \tag{2-93}$$

where m is the total mass of all rings and ω is the angular velocity.

The probability that the centrifugal force will exceed a specific value is equal to

$$P(P_c > P_c^*) = P\left(\rho_n > \rho_n^* = \frac{P_c^*}{m\omega^2}\right) = \int_{\rho_n^*}^{\infty} f(\rho_n)\, d\rho_n \tag{2-94}$$

Substituting Eq. (2-90) into Eq. (2-94) and integrating gives

$$P(P_c > P_c^*) = \exp\left(-\frac{\rho_n^{*2}}{2\sigma^2}\right) \tag{2-95}$$

Since the parameter σ is a function of component tolerances [see Eqs. (2-91) and (2-92)], this probability depends on assigned tolerances. The above formulas, Eqs. (2-90)–(2-95), can be applied to any cylindrically shaped assemblies. So the presence of unbalanced forces is a direct result of manufacturing tolerances. Here only the tolerances on the rotating parts were considered. However, the tolerances on the geometry of bearings, bearing housings, and housing bores contribute to the unbalanced forces and also to the misalignment of the rotating shaft. Some types of bearings may be very sensitive to such misalignment, so that a combination of a random degree of misalignment with a random amplitude of the unbalanced forces results in a random lifetime of a bearing. Consider a simple example of the effect of tolerances on misalignment.

Example 2-12 A shaft on two supports is shown in Fig. 2-26. The distance between the supports is L. Assume that the axis for the left bore in the housing is taken as a reference axis (which means that the left bore surface is taken as a datum). Then the position of the shaft axis with respect to the bore axis is determined by the eccentricities of the bearing outer ring, balls, inner ring, and shaft plus the magnitude of a clearance in the bearing (clearance between the balls and the rings after the bearing is mounted on the shaft). The latter, however, is neglected here for the sake of simplicity. Assume that the system of balls in a bearing can be treated as a ring, the tolerances of which are equal to those for an individual ball. Then the eccentricity vector for the left support $\bar{e}_\ell$ is determined by Eq. (2-79):

$$\bar{e}_\ell = \sum_{i=1}^{4} \bar{e}_i \tag{2-96}$$

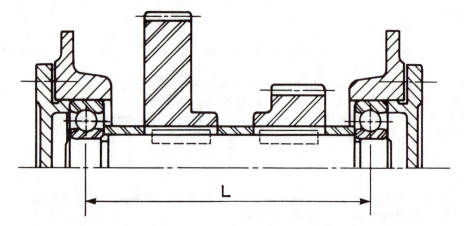

Figure 2-26 Gear-shaft assembly.

where $\bar{e}_1$, $\bar{e}_2$, $\bar{e}_3$, and $\bar{e}_4$ are the eccentricities for the outer ring, balls, inner ring, and shaft, respectively. Similarly, the eccentricity vector for the right support e_r is determined by

$$\bar{e}_r = \sum_{i=1}^{5} \bar{e}_i \qquad (2\text{-}97)$$

where $\bar{e}_1$, $\bar{e}_2$, $\bar{e}_3$, $\bar{e}_4$, and $\bar{e}_5$ are the eccentricities for the bore, bearing outer ring, balls, bearing inner ring, and shaft, respectively. (It is assumed here that the housings for two bearings in Fig. 2-26 cannot be bored with one setup, which means that the right bore may have a displaced geometrical axis with respect to the left). The statistical means for the eccentricity magnitudes in Eqs. (2-96) and (2-97) are related to the range of tolerances by Eq. (2-78), and the Rayleigh distribution parameters are found from Eq. (2-92). The resultant vector of eccentricities with respect to the datum is

$$\bar{e} = \bar{e}_\ell + \bar{e}_r \qquad (2\text{-}98)$$

The distribution of $|\bar{e}|$ is given by Eq. (2-90), in which ρ_n is substituted for e and

$$\sigma^2 = \sum_{i=1}^{9} \sigma_i^2 \qquad (2\text{-}99)$$

The probability that a misalignment angle $\alpha = e/L$ exceeds a specific magnitude $\alpha*$ is

$$P\,(\alpha > \alpha*) = \int_{e*}^{\infty} f(e)\,de \qquad (2\text{-}100)$$

where $e* = \alpha*L$.

As an illustration, consider a numerical solution. The following nominal dimensions are given: the shaft diameter at both supports is $d_s = 40$ mm, housing diameter at both supports is $d_h = 80$ mm. Assume that the same type of bearings is used at both supports, and neglect the effect of ball tolerances. Consider for simplicity a transition-type fit for both housing and shaft interfaces with the bearing rings. The following dimensions are assigned:

Shaft at two supports

$$d_s = 40^{+0.002}_{-0.014}$$

Bore of the bearing

$$d_b = 40^{+0.016}_{-0}$$

Outer diameter of the bearing

$$D_b = 80^{+0.003}_{-0.016}$$

Diameter of the housing

$$D_h = 80^{+0.019}_{-0}$$

Then the corresponding tolerances are

Shaft

$$\Delta_s = 0.016 \text{ mm}$$

Bore of the bearing

$$\Delta_{b1} = 0.016 \text{ mm}$$

Outer diameter of the bearing

$$\Delta_{b2} = 0.019 \text{ mm}$$

Diameter of the housing

$$\Delta_{h1} = 0.019 \text{ mm}$$

Also assume that the tolerance on the parallel misalignment of the right housing with respect to the left is $\Delta_m = 0.4$ mm. The calculation of the corresponding σ_j, according to Eq. (2-92), and σ, according Eq. (2-99), gives

$$\sigma^2 = \frac{1}{8\pi} (2 \times 0.016^2 + 2 \times 0.019^2 + 0.4^2) = 0.006415 \qquad \sigma = 0.08$$

For $L = 300$ mm the probability of exceeding various angles of misalignment is calculated using Eq. (2-100), which after integration yields an equation similar to Eq. (2-95):

$$P (\alpha > \alpha^*) = \exp \left(-\frac{e^{*2}}{2\sigma^2}\right) \tag{2-101}$$

The probability $P(\alpha > 0) = 1$, while the probability $P(\alpha > 0.001) = 0.00088$. In this particular case the probability of a significant misalignment is small.

Example 2-12 illustrates how the tolerances on various dimensions of different components in a system affect the probability of exceeding a specific angle of misalignment. However, the reliability of a bearing operating at a given speed and load under the conditions of misalignment is a property of the bearing itself. Thus the bearing reliability is conditional, the condition being in this case the probability of a specific position of the shaft. Note also that in this example two bearings are operating in similar conditions as far as misalignment is concerned. This may be a common cause for failure, other parameters being independent.

On Reliability of a Component in a System

As mentioned earlier, at the detailed design stage the reliability of a product, in principle, could be found if, in addition to the information about tolerances, the properties of the materials at various stress, temperature, and chemical environmental conditions were known, the internal and external environmental conditions could be predicted (in a statistical sense), and the external load conditions could be foreseen (either in a deterministic or statistical sense). Most often, not all of this information is available because the engineering science has not reached this stage of knowledge, engineering practice has not accumulated statistical data, or the product is a new design. However, even the mere identification of factors affecting component reliability, its source, and its importance is useful because adequate test programs can then be formulated, this information can be communicated to analytical engineers for computer simulation, or troubleshooting can begin for field failures. In this respect, consider the example of a lip-seal mentioned above.

Example 2-13 In Fig. 2-27 a single lip-seal separating a rotating shaft from the stationary case is shown. This type of a seal is usually used for retaining lubricants in machines having rotating and/or oscillating shafts. Various polymer materials are used for the seal elements. The function of the seal is determined by its ability to maintain a constant radial pressure between the sealing element and the shaft. Anything causing a decrease in this pressure affects leakage and thus seal failure. Consider factors affecting the radial lip-seal performance.

1. *Temperature*. Temperature affects oil viscosity and properties of the sealing material. If temperature increases, the viscosity of the oil goes down, and the oil film between the lip and the shaft becomes thinner. This leads to increased friction losses and thus to increased local temperature. The variation of temperature affects the sealing material as well: thermal expansion during temperature rise, and stress relaxation during temperature fall, speed up the process of material aging. Thus temperature control is important from the point of view of seal reliability. The temperature at the lip of the seal is affected by
   - ambient temperature
   - temperature of the oil
   - friction coefficient between seal and shaft
   - initial preload by a Garter spring
   - conductivity of materials.
2. *Wear*. Wear at the contact point of the seal directly affects its functioning. The rate of the wear depends on
   - coefficient of friction at the seal-shaft interface
   - pressure exerted by a Garter spring
   - pressure due to the seal-shaft fit
   - pressure caused by oil
   - shaft surface finish
   - shaft surface hardness
   - speed of rotation

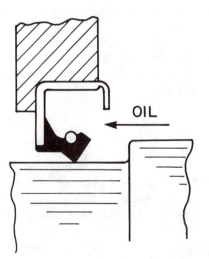

OIL

Figure 2-27 Lip-seal–shaft assembly.

- shaft oscillations
- abrasion resistance of the seal material.

3. *Fatigue*. Fatigue manifests itself in a gradual deterioration of the seal material, which loses its elastic properties. As a result, a seal cannot follow the shaft displacements, an event that constitutes a seal's failure. Factors affecting fatigue are
   - fluctuating temperature
   - chemical environment
   - oscillating shaft
   - pressure at the seal-shaft interface.

Not all of the above factors are equally important in seal reliability. To assign some weight to these factors would require more specific data about the design and the seal. Note that seal reliability is affected by the process of assembly as well, since it can simply be damaged while assembled. In practical situations, experience accumulated in industry is used as a guide.

Example 2-13 shows how complicated it is to assess the reliability of a component in an operating system. Again, there are factors associated with the component itself and with the system in which this component operates, and factors that are external to the system. Given the complexity of the physical processes and uncertainties associated with these processes and the system, true reliability is impossible to predict for a component in many situations previously unencountered. True reliability will reveal itself when the product is in service. In the meantime, at the detailed design stage, use of design experience and analytical and experimental methods will increase the relative level of reliability.

Thus over the entire design process, reliability constitutes one of the basic considerations for making decisions, and although the absolute level of reliability may not be known, its relative level is growing during this process.

PROBLEMS

2-1 A brake for a hoist is shown in Fig. P2-1. Two springs are exerting forces on the corresponding shoes, thus keeping the brake normally locked. To unlock the brake, a hydraulic actuator is used.

(a) Draw a reliability diagram of the brake considering only components indicated.

(b) Draw a reliability diagram if the brake is normally unlocked but becomes locked when hydraulic pressure is decreased. Explain the difference between the two cases.

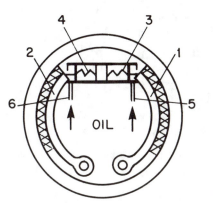

Figure P2-1

2-2 Consider a tooth in a gear. Is it in series or in parallel with other teeth? Is a spline in series or in parallel with other splines?

2-3 Consider the reliability of a tire on a four-wheel vehicle. Discuss the effect of the tire on system reliability and the effect of the system on tire reliability. What is the notion of an interface in this case?

2-4 A sheave with three V-belts is shown in Fig. P2-4. Identify factors affecting the reliability of the belts and the reliability of the sheave-belt interface. Draw a reliability diagram of the drive in which blocks for belts and interfaces are identified.

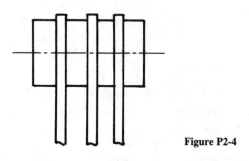

Figure P2-4

2-5 Three possible configurations of gear reduction boxes are shown in Fig. P2-5. Assuming that the input-output characteristics of all boxes are identical, (*a*) discuss the reliability advantages and disadvantages of every configuration, and (*b*) draw a reliability diagram for each configuration, considering only such components as shafts, gears, and bearings.

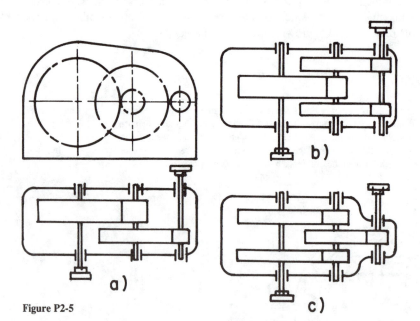

Figure P2-5

2-6 Two possible configurations of a spur gear–support system are shown in Fig. P2-6.
    (*a*) Discuss the reliability advantages and disadvantages of both configurations.
    (*b*) Draw a reliability diagram of the gear-bearings-shaft system.
    (*c*) Discuss the influence of the system on the bearings' reliability.

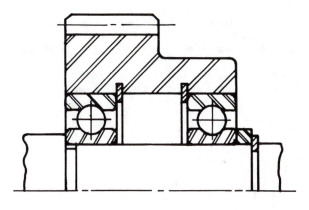

a)

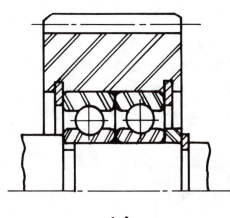

b)

Figure P2-6

2-7 Compare reliabilities of the configurations in Figs. P2-5a and P2-5b. Assume that the reliability functions for shafts and gears remain the same in both configurations, whereas the reliability function for the bearing is expressed as $R(t) = \exp(-ht)$, where $h = \alpha P^\beta$ $\alpha > 0$, $\beta \geq 1$, and P is the magnitude of force acting on the bearing. Assume that forces acting on bearings 3 and 4 are the same in both configurations, whereas forces acting on other bearings are as follows:

Configuration a

$$P_{1a} = \tfrac{1}{3}P_1 \qquad P_{2a} = \tfrac{2}{3}P_1$$

$$P_{5a} = \tfrac{2}{3}P_3 \qquad P_{6a} = \tfrac{1}{3}P_3$$

Configuration b

$$P_{1b} = P_{2b} = \tfrac{1}{2}P_1$$

$$P_{5b} = P_{6b} = \tfrac{1}{2}P_3$$

where P_1 and P_3 are the total forces acting on the input and output shafts, respectively.

(a) Determine the critical value of β when configuration b becomes more reliable than a.

(b) Discuss other factors affecting the reliabilities of two configurations.

2-8 An electrical motor and a pump are coupled together by a clutch and installed on a common foundation (see Fig. P2-8).

(*a*) What additional factors may affect the reliability of the motor and pump after they were joined together?

(*b*) Is it possible that some failures in the motor and pump would be statistically dependent?

(*c*) What is the effect of the assumption of statistical independence on system reliability?

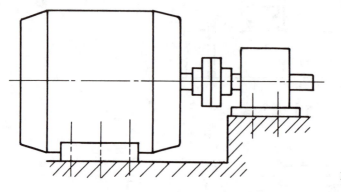

Figure P2-8

2-9 In Fig. P2-9, three alternatives of power-generating units are shown, where b denotes a boiler and t a turbine.

(*a*) Assuming that reliabilities of boilers and turbines in all configurations are the same, which alternative is more reliable?

(*b*) Indicate degrees and levels of redundancy in these alternatives.

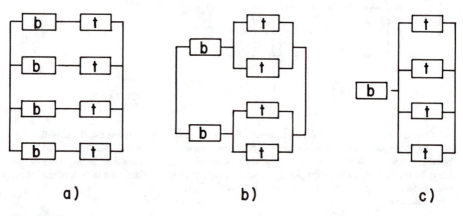

a) b) c)

Figure P2-9

2-10 Consider a gas pumping station with two compressors. If one compressor is sufficient to supply the gas, then two alternatives are possible: (1) one compressor is operating at full capacity, while the other is in standby mode, and (2) two compressors are operating simultaneously. In both cases the compressors are in a random failure mode; however, in the first case the failure rate is two failures per year, whereas in the second case it is 1.5 failures per year when two are operating and two failures per year when only one is operating. Find reliability functions for both cases and determine which alternative is more reliable.

2-11 A flexible coupling can be designed with four, six, or eight torque-transmitting elastic elements (see Fig. P2–11). Four is the minimum number of elements needed to transmit the torque. Calculate the reliability of the three coupling configurations if the reliability of the elastic element after 1 year of operation is (a) $R = 1 - \exp(-3.21)$ and is independent of the load on the element, and (b) $R = 1 - \exp(-3.21/\bar{F})$, where

$$\bar{F} = \frac{\text{force on the element for } N > 4}{\text{force on the element for } N = 4}$$

N is the number of elastic elements in a coupling; thus, in this case the reliability is load-dependent. Which approach is a more conservative one?

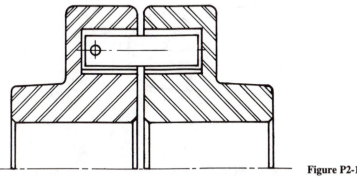

Figure P2-11

2-12 Two concentric springs make up a support (see Fig. P2-12). Both springs are known to be in a state of random failure with the rates $h_1 = 0.005$ failures per kilocycle and $h_2 = 0.008$ failures per kilocycle for the inner and outer springs, respectively. Each of two springs is capable of withstanding the fluctuating load. However, if the inner spring fails, the failure rate of the outer one becomes $h_2^* = 0.012$ failures per kilocycle; if the outer spring fails, the failure rate of the inner one becomes $h_1^* = 0.0075$ failures per kilocycle. What is the reliability function of the support from 0 to 400 kilocycles?

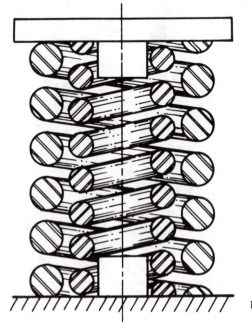

Figure P2-12

2-13 In problem 2-12, assume that the failure rates under the increased load when one of two springs fails remains the same (the static part of the increased load does not affect the failure rate). How does this assumption affect the reliability function in the range 0–400 kilocycles?

2-14 Consider a belt drive (Fig. P2-14) designed to transmit a constant torque with constant speed. In reality, the power transmitted by the belt is a random variable.

(a) Identify parameters of a belt drive affecting the randomness of the power transmission.

(b) Discuss the conditions when predicting belt performance is a problem of probabilistic design and when it becomes a problem of reliability engineering.

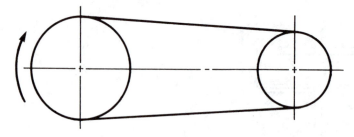

Figure P2-14

2-15 The torque transmitted by a belt drive is proportional to the tension F_i (see Fig. P2-15), i.e., $M_a = AF_i$, where M_a is the torque and A is the coefficient of proportionality. Because of the relaxation of the belt, this tension can be considered as decreasing in time according to the equation

$$F_i = F_{i0} \exp\left(-\frac{t}{t_R}\right)$$

where t_R is relaxation time and F_{i0} is initial tension.

(a) If F_{i0} is normally distributed with $\mu_F = 100N$ and $\sigma_F = 10N$, what would the reliability of the belt drive be in the case of $t_R = 100$ hours, $A = 0.25$ m, and external torque $M_e = 20$ N m?

(b) How does the safety factor change in time?

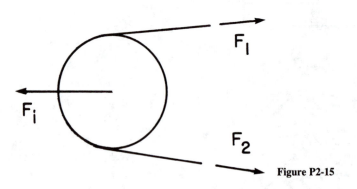

Figure P2-15

2-16 Consider problem 2–15 again and assume that the external torque is also a random normally distributed variable with $\mu_e = 20$ N m and $\sigma_e = 2$ N m.

(a) How does the reliability function change?

(b) Does the variability of the external torque affect the safety factor?

2-17 It is found that in a rotating beam made out of carbon steel the maximum completely reversed stress is normally distributed with the parameters $\mu_\sigma = 350$ MPa and $\sigma_\sigma = 30$ MPa. The fatigue strength is known to be within the bounds shown in Fig. P2-17. Assume a normal distribution for the fatigue strength and find the parameters of the distribution using the 3-σ rule for any given value of N. What is the probability of survival (reliability) at 10^4 cycles, 10^5 cycles, and 10^6 cycles?

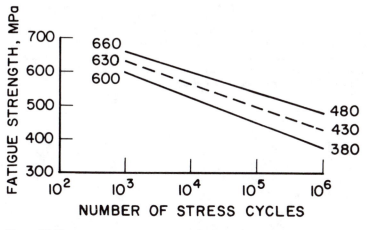

Figure P2-17

2-18 (a) Find the reliability of operating for $N = 10^6$ cycles in problem 2-17 for three different stress variability levels: $V_\sigma = 0.1, 0.15,$ and 0.20.

(b) Compare the reliability with the safety factor to operate 10^6 cycles for the data given in problem 2-17.

2-19 Three gears are shrink fitted on a shaft as shown in Fig. P2-19.

(a) Find the probability that the unbalanced force from each gear exceeds 1 kN and 3kN.

(b) Find the probability that the total unbalanced force exceeds 1 kN and 3 kN.

(Assume that the gears are made out of steel, that the shaft rotates 3000 rpm, and that the shaft's center of mass coincides with its axis of rotation.)

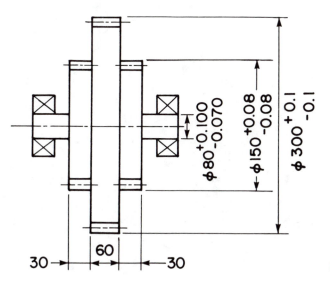

Figure P2-19

2-20 Two perfectly balanced shafts are rigidly joined together. If the allowable misalignment is $\Delta = 2$ mm, the masses of shafts are $m_1 = 200$ kg and $m_2 = 400$ kg and their centers of gravity are 2 m apart, what is the probability of exceeding a 10 kN centrifugal force and 10 kN m centrifugal bending moment at 3000 rpm? Assume that Δ is Rayleigh distributed with the parameter

$$\sigma = \left(\frac{2}{\pi}\right)^{1/2} \frac{\Delta}{2}.$$

2-21 Clearance between a turbine blade and a stator is specified to be $\Delta = 1$ mm. The nominal length of the blade is 300 mm, and it is known to be creeping at elevated temperatures according to the formula $d\varepsilon/dt = 10^{-7}$ h^{-1}, where $d\varepsilon/dt$ is the strain rate. If the clearance is normally distributed and a zero clearance constitutes a failure, find the reliability of the rotor-stator interface.

THREE

TESTING FOR RELIABILITY

3-1 INTRODUCTION

The reliability of a component or interface in a system very often depends on many interrelated parameters, each of which has a margin of uncertainty (see Chapter 2). If the effect of various uncertainties cannot be quantified by other means (analytical techniques, computer simulation, previous experience), then testing is in order. The need for a test comes after a design engineer does a thorough analysis of factors affecting product reliability and identifies critical factors. As a result of a designer's assessment of potential problems, it should become clear what kind of objects to test (components, subsystems, product), the conditions for testing (parameters and their ranges), and the time when the test results are needed.

To understand the problem facing a design engineer, one must look again at the design process. The design process is a series of decisions based on the state of knowledge and available information. Any test result is needed only as input information in this decision-making process and not for itself (only tests tied to a specific design are assumed here). So the testing process must run ahead of the design process if no delays are expected. Note that the question of integrating the testing process into the design process is not of concern here, since it is a problem of project management; the point here is that if extensive and time-consuming tests are to be done, then the decision about these tests should be made well in advance. A clear conflict is seen here: a decision about the testing of a design should be done before this design exists. The probability of making a wrong decision (to test or not to test) is conditional upon a designer's experience.

3-2 TYPES OF TESTS

The decisions about what kind of objects to test are made as early as the conceptual stage. At this stage, components critical to the concept validity are identified and the conditions these components must withstand are specified. The planning, execution, and analysis of test results should be done before or at the early steps of the preliminary design stage. A blade of a steam turbine (Example 1–3) is an example of a component whose criticality becomes clear at the conceptual stage, when the appropriate power-speed-temperature specifications are outlined. At this time the designer will specify the speeds of rotation and the temperatures at which the blade will be tested. Then blades made of various materials and by various technological methods are tested.

The integration of a component into a system can be done in various ways. The reliability considerations (among other considerations such as assembly and repair) may significantly affect the specifications for manufacturing a component due to the interaction of this component with a system. So before these specifications are outlined at the detailed design stage, the effect of the system on a component should be known. By virtue of timing, the experimental investigation should be done at the preliminary design stage. Consider again an example of the blade. In the operating turbine a blade is not only subjected to the steady centrifugal forces but also vibrates due to steam-blade and rotor-blade interactions. In the presence of vibrations the rate of blade creep may increase and thus the probability of survival for a specified period of time will decrease. The effect of various levels of vibrations and design solutions on the rate of creep should be investigated at the preliminary design stage and implemented at the detailed design stage.

After the detailed design stage is completed and before a production process starts, usually a prototype of the product is built and tested. This completes the test program during the product development phase.

The characteristic feature of the reliability test program is its shift in phase and, correspondingly, in emphasis with respect to a design process: from components at the conceptual stage, to a subsystem at the preliminary stage, and to a system at the detailed stage.

At this moment it should be mentioned that there are tests with other objectives besides reliability that are part of the product's development: the development, qualification, and acceptance tests. The development tests are done in order to verify or develop a new concept. The qualification tests are done in order to verify that the product meets design requirements and thus can be qualified for its intended application. The acceptance tests are done in order to verify that a particular product (or batch) meets the operating requirements prior to acceptance by the user. It is seen that all of these tests are functional tests, i.e., the objective of these tests is to verify the quality of performance under simulated environmental and operational conditions, while the duration of the test is not important. Reliability is also concerned with the quality of performance; however, the emphasis is on how this quality changes in time. It is clear that functional tests can increase or decrease the designer's confidence in the product; however, they cannot substitute the reliability tests. The following example illustrates this point.

Example 3-1 Consider a newly designed hydraulic transmission that is supposed to be used for greater loads than before and discuss what is involved in the development, qualification, and acceptance tests.

The development test would be performed on a prototype unit to check whether indeed the maximum torque at the maximum speed can be achieved. The objective thus is very limited.

The qualification test would be done to check whether kinematic, dynamic, efficiency, noise, and vibration parameters are within the design limits for this transmission. This test is done on a prototype unit by simulating load and environmental conditions.

The acceptance test would be done for each unit (or for a sample of units), as part of the quality control system, after manufacture and assembly. The test may include an idle run, backlash measurements, test of seals under pressure, etc.

Basically, all these tests would be done to prove that the transmission operates at the speeds and loads specified.

Three distinctive differences between reliability and functional tests are important.

1. The functional tests are done when a prototype is built, i.e., at the end of the design development process, whereas reliability tests start as early as the conceptual stage of the design process.
2. The functional tests are not time consuming in comparison with reliability tests, since in the latter there is a need to bring an item to the state of failure, whether it is wear, fatigue, or corrosion.
3. After completion of functional tests, an item can be used again, whereas after reliability tests, an item will either be discarded or need repair. This factor affects the costs of the test.

The reliability tests may be done with two different objectives: to verify a specific design with fixed parameters, and to optimize the reliability by finding experimentally the best set of parameters. For example, in a sliding bearing, if the geometry, oil viscosity, and load are fixed, then what is needed is to test a sample of bearings in order to obtain statistics about failure times. However, if the oil viscosity and bearing geometry (clearance) are variables, then the objective is to find the best combination that results in maximum reliability.

Two different objectives lead to two different methodologies of carrying out the tests. The reliability verification test is straightforward, whereas the optimization tests require a special methodology. For the case of a sliding bearing the optimization test could be carried out by making all possible combinations of viscosity and clearance and finding the corresponding reliabilities. Not only is such a procedure time consuming, but it can also produce erroneous results. A completely randomized procedure, one sample with the random combination of values of two variables, requires less time and gives a desired degree of confidence.

Another distinctive feature of reliability tests is their statistical nature, that is, the test results have a specified degree of confidence that depends on the sample size. Thus the sample size and the way a sample is drawn from the population are very important. The procedure for identification of variables in experiments, the selection of items to be tested, the decision on sample size, and the analysis of experimental data constitute the test methodology, called the statistical design of experiments.

A combination of a multivariable test, a complicated test procedure, and the statistical nature of the test can make the problem of reliability estimation by testing very expensive and time consuming. In this respect a decision about what kind of objects (components or products) to test becomes very important. Let us consider briefly the advantages and disadvantages of testing components versus testing products.

The testing of components has the following advantages.

1. Methodological, since the number of variables is smaller
2. Timing, since test results can be implemented before the product is built
3. Costs and time consumption.

The disadvantages of testing components only are

1. Incompleteness, since common cause failures and the effect of parameters that are the result of system functioning cannot be estimated
2. Diversity, i.e., a possible need to test various components under various conditions.

3-3 DESIGN OF EXPERIMENTS

The uncertain elements in the design should be identified early on, and corresponding parts and interfaces tested. However, because of the time needed to conduct a reliability test, to integrate the test into a design process, and the costs involved, usually only those tests are planned whose criticality is in no doubt and in cases when reliability requirements are absolutely important (as in the case of safety). If a decision to conduct an experimental investigation is made, the investigation should be properly designed and executed, and the results analyzed. In the following example the properties of a shrink-fit joint are considered as they relate to joint reliability, and the elements of the methodology on the statistical design of experiments are demonstrated.

Example 3-2 A shrink-fit joint required by a particular design was too long to meet the size constraint on the entire product. The bushing-shaft interference could not be increased any more because of the magnitude of contact stresses. So the design engineer, after studying various alternatives of increasing the coefficient of friction at the interface, came across an innovative idea to introduce a tiny film between the two mating parts by electroplating the surface of one of the parts.

The effect of the film on the coefficient of friction stems from the following considerations. A coefficient of friction depends on the true area of contact of two mating parts and increases with the increase of this area. The true area of contact depends on the contact pressure and hardness of surfaces of two mating parts, but to a greater degree, it depends on the geometry of surfaces: roundness and straightness of both shaft and hole and surface roughness consisting of fine irregularities superimposed on wavelike variations. A tiny film of soft material (such as copper or zinc) comparable in thickness with the height of asperities would fill the spaces between the two mating parts, thus increasing its true contact area and, as a result, increasing the coefficient of friction. Then the length of the joint interface could be shortened, and the design would meet the size constraint. The effectiveness of coated joints had to be estimated before the preliminary design advances. A test would consist of finding a static coefficient of friction when an axial force is applied to the shaft and it starts sliding with respect to the bushing.

Before starting test preparations, the design engineer should formulate exactly what he or she wants to achieve. In the case under consideration the objective is to find out the effect of the copper film on the coefficient of friction in a shrink-fit joint, and also the effect of the film thickness. To fulfill this objective may not be a simple task. First, the coefficient of friction depends on other parameters, such as parts geometry, surface finish, surface hardness, surface contamination, and contact pressure, and as a result, there is a problem of isolating the effect of the film on the coefficient of friction. In addition, all parameters are random, so that the magnitude of each cannot be controlled during the experiment. The problem is how to conduct experiments to ensure with a specified degree of confidence the validity of the results while minimizing the number of test runs.

The methodology of solving the above problem is developed in statistics and is called the design of experiments. A good understanding of concepts involved in designing experiments is needed on the

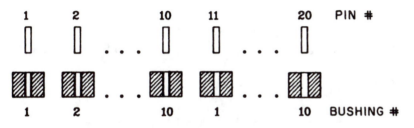

Figure 3-1 Arrangement of pin-bushing couples without randomization.

part of a design engineer in order to efficiently participate in the planning of tests and interpretation of results.

Let us assume that only the effect of film thickness is to be investigated. Three different approaches to the planning and conducting of tests can be developed. Let us assume that 10 specimens without and 10 with plating are to be tested. Then 20 pins and 20 bushings are manufactured, and the inner surface of 10 bushings is electroplated with a film of a specific thickness (note that the film thickness is also a random variable, but it can be assumed that the mean of it remains the same for all bushings).

One way of conducting the experiments would be to assign numbers to all pins (from 1 to 20), to all bushings without plating (from 1 to 10), and to all bushings with plating (from 1 to 10), then to assemble first pin with first bushing with a copper film, second pin with second bushing, and so on, then eleventh pin with first bushing without a copper film, twelfth pin with second bushing, etc. (see Fig. 3-1). This test will be called a test without randomization. What is the drawback of this arrangement? Assume that all pins and bushings are made in accordance with the design specifications on the geometry and surface finish. Even if they all were machined on the same lathe, a trend in machine variability may introduce a trend in joint properties. In statistical terminology it means that the above ordering of specimens introduces a bias into the inference about the fit properties.

To eliminate the effect of bias, the experiment may be designed differently. Pin and bushing numbers can be allocated in a random fashion. That is, out of 20 numbers a random number is generated (e.g., using tables) and a pin with this number is chosen (say, 11), then another random number out of 20 (all bushings are now numbered as one set) is generated, and a bushing with this number is chosen (say, 2). Then, a couple (11, 2) becomes a first couple to be tested. In this way, 20 random couples are generated (see Fig. 3-2 and Table 3-2), but only the first 10 couples are made with the film at the interface. This process of arranging specimens in a test set is called randomization. Its intention is to neutralize on average the effect of a trend in pin and bushing geometries on fit properties. This test will be called a test with randomization.

Now, let us assume that 20 tests of fits were done, and the distributions of friction coefficients for fits with and without copper film were found (see Fig. 3-3, where μ_{f1}^* and μ_{f2}^* are the sample mean friction coefficients for a fit without film and with film, respectively). It is seen that $\mu_{f2}^* > \mu_{f1}^*$ and the positive effect of the copper film is thus proven. However, the question of validity of test results is still not certain owing to variance in the amount of metal interference between any assembly with and without film in Fig. 3-2, which is to say that a difference in the coefficient of friction may not only be due to the effect of the film but also to the difference in contact pressure. In fact, instead of having one

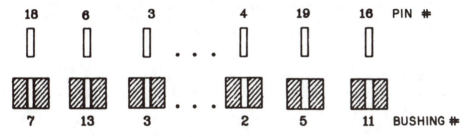

Figure 3-2 Arrangement of pin-bushing couples with randomization.

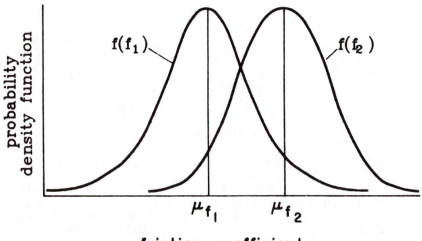

Figure 3-3 Distributions of coefficients of friction for fits without (f_1) and with (f_2) copper film.

variable parameter, film, there are two variable parameters, and the second one, contact pressure, remains uncontrolled. To eliminate this flaw, the experiment should be redesigned.

Now instead of 20 pins and 20 bushings, only 10 of each will be manufactured. However, each pin and bushing will be at least 2 times longer. As before, a set of random couples (pin, bushing) will be generated (see Fig. 3-4). For each couple it can be assumed that the metal interference remains the same (on average) along the length of the interface. Now let us cut each bushing into two halves, electroplate one of these halves with a copper film, and shrink-fit both of them on the same pin. If forces are correctly applied, each bushing on the same pin can be independently tested and the corresponding friction coefficient found. In this test the cause of variance between a fit with and without film is the film itself, other factors being statistically neutralized. This method (pairing of two fits with and without film such that within a pair there are no factors, other than the film, statistically affecting the friction) is called blocking. In Fig. 3-4, pairs (10, 3), (6, 5), . . . are blocks. Randomization and blocking are two distinct principles of design of experiment. The third test will be done with both randomization and blocking.

Let us demonstrate the effect of randomization and blocking on test outcome. For a shrink-fit the following dimensions for pin and hole diameters were specified: pin diameter $20^{+0.062}_{-0.041}$, hole diameter $20^{+0.021}_{-0}$ (dimensions in millimeters). To conduct the experiment according to Fig. 3-1, 20 pins and 20 bushings were manufactured. Each pin and bushing was numbered as it was manufactured, and the dimensions for each were measured. The results are shown in Fig. 3-5.

Three approaches to the designing of experiments are demonstrated below.

First experiment. In the first nonrandomized experiment the pin-bushing pairs were formed as follows: (1, 1), (2, 2), . . . ,(20, 20), and the first 10 pairs had a 4-μm film coating. The test sequence, the corresponding pairs, the metal interference, and the coefficient of friction are shown in Table 3-1. The average of friction coefficients with and without the coating based on the data in Table 3-1 is $\mu^*_{f1} = 0.204$ and $\mu^*_{f2} = 0.118$, respectively. Tests seem to show a 2-to-1 advantage to using 4-μm copper film. However, it is important to check whether this conclusion is statistically sound by using methods of statistical hypothesis testing.

The uncertainty of the inference based on the limited experimental data is associated with the variance of the data. In this situation the mean itself is a random variable whose variance equals

$$\sigma^2_{\mu f} = \frac{\sigma^2}{n} \tag{3-1}$$

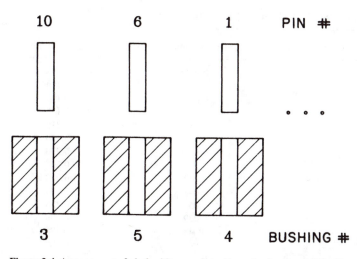

Figure 3-4 Arrangement of pin-bushing couples with randomization and blocking.

where σ^2 is the variance of the population and n is the sample size.

It is seen from Eq. (3-1) that for a reasonably large sample size the variance of the measured response (in the example above it is the coefficient of friction) tends to zero, which means that the estimates of averages would be sufficient to decide the effect of the copper film on friction. However, for a small sample size the variance of the average cannot be neglected. The statistical method, which compares two (or more) populations in terms of the parameters of these populations, is called the analysis of variance (ANOVA) method. In order to see the application of the ANOVA method to our problem in Example 3-2, a brief outline of the method is given first.

The first step in the ANOVA method is the formulation of the statistical problem. In our case it is of interest to assess whether a conclusion that the film has increased

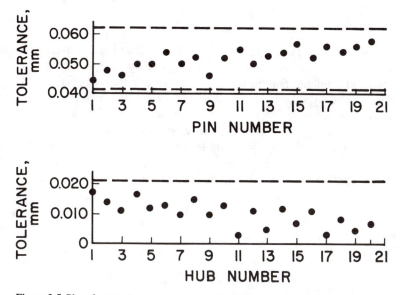

Figure 3-5 Plot of successive pieces of pins and bushings from a manufacturing process.

Table 3-1 Nonrandomized Test Data and Results

Test number	Pin-bushing pair	Metal interference	Coefficient of friction
1	1, 1	0.031*	0.155
2	2, 2	0.038*	0.190
3	3, 3	0.039*	0.195
4	4, 4	0.037*	0.185
5	5, 5	0.042*	0.210
6	6, 6	0.048*	0.240
7	7, 7	0.044*	0.220
8	8, 8	0.041*	0.205
9	9, 9	0.040*	0.200
10	10, 10	0.047*	0.235
11	11, 11	0.051	0.128
12	12, 12	0.039	0.098
13	13, 13	0.048	0.120
14	14, 14	0.042	0.105
15	15, 15	0.050	0.125
16	16, 16	0.041	0.103
17	17, 17	0.053	0.133
18	18, 18	0.045	0.113
19	19, 19	0.051	0.128
20	20, 20	0.051	0.128

* Here, 4-μm film thickness is included.

the friction coefficient twice is statistically justified or, alternatively, whether it is not true. These two statements are called the hypotheses, and their verification is called hypothesis testing. Formally, two hypotheses are written in the form

$$H_0: \mu_{f1} - \mu_{f2} = \gamma \tag{3-2}$$

$$H_A: \mu_{f1} - \mu_{f2} \neq \gamma \tag{3-3}$$

where μ_{f1} and μ_{f2} are the means of the first and second populations, respectively, and γ is the specified difference between the two means. The first hypothesis H_0 is called the null hypothesis and it implies that the indicated difference in the means is statistically correct. The second hypothesis H_A is called the alternative hypothesis, and it implies that the H_0 hypothesis is incorrect. It is clear that the two hypotheses are mutually exclusive, i.e., if H_0 is true, then H_A is false, and vice versa.

In our case the fact that $\mu_{f1} - \mu_{f2} \neq \gamma$ may not satisfy the design engineer, who may wish to know whether $\mu_{f1} - \mu_{f2} > \gamma$ or $\mu_{f1} - \mu_{f2} < \gamma$, which means that two different alternative hypotheses can be formulated for the same null hypothesis:

$$H_0: \mu_{f1} - \mu_{f2} = \gamma \tag{3-4}$$

$$H_{A1}: \mu_{f1} - \mu_{f2} < \gamma \tag{3-5}$$

$$H_{A2}: \mu_{f1} - \mu_{f2} > \gamma \tag{3-6}$$

The second step in the ANOVA method is the calculation of the sample statistics—means and variances—and deciding on an appropriate model for the sampling distribution. The sample mean is

$$\mu_f^* = \left(\sum_{i=1}^{n} f_i \right) n^{-1} \tag{3-7}$$

and the sample variance is

$$S^2 = \left[\sum_{i=1}^{n} (f_i - \mu_f^*)^2 \right] (n - 1)^{-1} \tag{3-8}$$

If it can be assumed that the population has a normal distribution of friction coefficients, then the random variable t having the Student's t-distribution is the test statistic.

In our case the test statistic is

$$t = \frac{\mu_{f1}^* - \mu_{f2}^* - \gamma}{S_p(1/n_1 + 1/n_2)^{0.5}} \tag{3-9}$$

where

$$S_p^2 = \frac{(n_1 - 1)S_1^2 + (n_2 - 1)S_2^2}{n_1 + n_2 - 2} \tag{3-10}$$

and S_1 and S_2 are the sample variances with and without the copper film, n_1 and n_2 are the sample sizes for the two types of fits with and without the film, S_p is called the pooled variance, and γ is the constant value by which μ_{f1} and μ_{f2} differ.

The third step is choosing the level of confidence, which is associated with the decision based on the given statistical data. Since the means μ_{f1} and μ_{f2} are random variables, there is a chance that the range of estimates of these means based on samples will not include the true mean. The probability that the true mean is included in the interval

$$\mu_{fL} \leq \mu_f \leq \mu_{fU} \tag{3-11}$$

is stated as

$$P(\mu_{fL} \leq \mu_f \leq \mu_{fU}) = 1 - \alpha \tag{3-12}$$

Then the interval Eq. (3-11) is called $100(1 - \alpha)$ percent confidence interval for the parameter μ_f. The statistics μ_{fL} and μ_{fU} are the lower and upper confidence limits, respectively, $1 - \alpha$ is the confidence coefficient, and α is the level of significance. If $\alpha = .05$, the interval Eq. (3-11) is a 95 percent confidence interval for μ_f. Usually tables are given for 95 and 99 percent confidence intervals.

The fourth step is finding the critical values of the test statistic (t). Critical values separate the region of rejection from the region of acceptance on the distribution of the test statistic. The region of rejection contains those values of t that would have a small probability of occurrence if the null hypothesis was true. Correspondingly, the region of acceptance contains those values of t that would be most probable if the null hypothesis was true. In our case, for the null hypothesis given by Eq. (3-2), the regions of rejection corresponding to the alternative hypotheses Eqs. (3-3), (3-5), and (3-6) are given by

$$\text{If } H_A: \quad \mu_{f1} - \mu_{f2} \neq \gamma \quad \text{then } t < -t_{\alpha/2,\nu} \quad \text{and } t > t_{\alpha/2,\nu} \quad (3\text{-}13)$$

$$\text{If } H_{A1}: \quad \mu_{f1} - \mu_{f2} < \gamma \quad \text{then } t < -t_{\alpha,\nu} \quad (3\text{-}14)$$

$$\text{If } H_{A2}: \quad \mu_{f1} - \mu_{f2} > \gamma \quad \text{then } t > t_{\alpha,\nu} \quad (3\text{-}15)$$

where $t_{\alpha/2,\nu}$ and $t_{\alpha,\nu}$ are the critical values of t and $\nu = n_1 + n_2 - 2$ is the number of degrees of freedom.

It is seen from conditions Eqs. (3-13)–(3-15) that the critical value of the test statistic depends on the hypothesis statement, the confidence level, and sample size.

Example 3-2 (Continued) In the case of the first experiment, $n_1 = n_2 = 10$, so that the number of degrees of freedom $\nu = 18$. Let us assume a 95 percent confidence interval, which means that $\alpha = .05$. For given α and ν, the critical values of the test statistic are found from the t-distribution table (see Appendix B): $t_{\alpha/2,\nu} = 2.101$ and $t_{\alpha,\nu} = 1.734$. The sample variances according to Eq. (3-8) and data in Table 3-1 are $S_1^2 = 0.000623$ and $S_2^2 = 0.000155$ for the samples with and without film, respectively. The pooled variance is $S_p = 0.0197$. Let us take $\gamma = \mu_{f2}$, which means that the null hypothesis to be tested is

$$H_0: \quad \mu_{f1} = 2\mu_{f2}.$$

For the given data the test statistic is $t = -3.63$. The null hypothesis will be rejected if $|t| > t_{\alpha,\nu} = 2.101$. So indeed the null hypothesis should not be accepted, which means that there is no sufficient statistical evidence to conclude that the average friction coefficient is 2 times greater in a fit with a copper film than in a fit without the film.

Table 3-2 Randomized Test Data and Results

Test number	Pin-bushing pair	Metal interference	Coefficient of friction
1	18, 7	0.048*	0.240
2	6, 13	0.053*	0.265
3	3, 3	0.039*	0.195
4	2, 8	0.037*	0.185
5	5, 20	0.047*	0.235
6	7, 6	0.041*	0.205
7	9, 4	0.037*	0.185
8	8, 9	0.046*	0.230
9	14, 16	0.047*	0.235
10	1, 14	0.036*	0.180
11	4, 2	0.036	0.090
12	19, 5	0.044	0.110
13	17, 19	0.051	0.128
14	11, 1	0.038	0.095
15	12, 17	0.047	0.118
16	20, 10	0.045	0.113
17	15, 15	0.050	0.125
18	10, 18	0.043	0.108
19	13, 12	0.042	0.105
20	16, 11	0.049	0.123

* Here, 4-μm film thickness is included.

Table 3-3 Randomized Test Data with Blocking

Pin-Bushing Pairs

	Block									
	1	2	3	4	5	6	7	8	9	10
Film	3, 2	9, 5	4, 8	6, 4	8, 3	7, 7	2, 1	1, 6	5, 9	10, 10
No film	3, 2	9, 5	4, 8	6, 4	8, 3	7, 7	2, 1	1, 6	5, 9	10, 10

Second experiment. In the second experiment, instead of ordering pin-bushing couples like those in the first test (see Table 3-1), the randomized ordering is arranged and the results are given in Table 3-2. The sample mean friction coefficients and the variances with and without the film are $\mu_{f1}^* = 0.215$, $\mu_{f2}^* = 0.111$, $S_1^2 = 0.000853$, and $S_2^2 = 0.000158$, respectively. The pooled variance is $S_p = 0.0225$. The test statistic in the case of $\gamma = \mu_{f2}^*$ and after substitution of the corresponding values in Eq. (3-9) is $t = -0.70$. It is seen that since α and ν are the same as in the previous case, the critical values of the test statistic are also the same and thus, since $|t| < t_{\alpha/2,\nu} = 2.101$, the null hypothesis should be accepted.

Third experiment. In the third experiment the blocking of pins and bushings was arranged (see Fig. 3-4) in such a way that the two pairs in one block, with and without a film, cannot now be considered as being independent. Instead of $2n - 2$ degrees of freedom, there are now only $n - 1$ degrees of freedom. The pin-bushing pairs were randomly selected (see Table 3-3), and the corresponding coefficients of frictions were found (see Table 3-4). The difference d between the two friction coefficients in a pair is shown in Table 3-4. The sample mean difference is $\mu_d^* = 0.11$; let us formulate the test hypothesis for the difference as follows

$$H_0: \quad \mu_d = \mu_0 \tag{3-16}$$

$$H_A: \quad \mu_d \neq \mu_0 \tag{3-17}$$

where μ_0 is found from the same requirement that $\mu_{f1} = 2\mu_{f2}$, which, since $\mu_d = \mu_{f1} - \mu_{f2}$, gives $\mu_d = \mu_{f2} = \mu_0$. Note that $\mu_{f1}^* = 0.2$ and $\mu_{f2}^* = 0.09$ are the sample means for the first and second rows in Table 3-4, and thus $\mu_0 = \mu_{f2}^* = 0.09$.

The test statistic is

$$t = \frac{\mu_d - \mu_0}{S/\sqrt{n}} \tag{3-18}$$

The null hypothesis will be rejected if $|t| > t_{\alpha/2,\nu}$, where $\nu = n - 1$. For $\alpha = 0.05$ and $n = 10$ the critical value of the test statistic is $t_{0.025,9} = 2.262$, whereas according to Eq. (3-18), in which the variance $S = 0.0098$, the test statistic is $t = 8.31$.

Table 3-4 Test Results

Coefficients of Friction

	Block									
	1	2	3	4	5	6	7	8	9	10
Film	0.180	0.19	0.195	0.205	0.225	0.220	0.175	0.175	0.220	0.215
No film	0.080	0.085	0.088	0.093	0.103	0.100	0.078	0.078	0.100	0.098
Difference	0.1	0.105	0.107	0.112	0.122	0.120	0.097	0.097	0.120	0.117

Thus the null hypothesis should be rejected, which means that there is no statistical evidence to suggest a 95 percent confidence that the friction will be twice as large in a coated joint as in a noncoated one.

A few comments should be made with respect to the above example.

1. Differently designed experiments produce different results for the averages.
2. The test results in all three experiments indicate that the friction coefficient in a joint with a copper film is practically 2 times greater. However, only the second statistical hypothesis test supports this assumption. The explanation of discrepancy is associated with the very small variances of data in all three tests. Let us imagine as an extreme case that the variances tend to zero. Then, for example, μ_d and μ_0 in Eq. (3-16) will be two distinct deterministic numbers, and although the difference between the two may be negligibly small, since it is not zero, the null hypothesis will be rejected. In other words, the hypothesis for small variances becomes sensitive to small variations in data. From the practical point of view the sensitivity of results with respect to the hypotheses formulated is important. In this respect the conclusion of the statistical analysis should be weighed against practical considerations.
3. A design engineer should understand the procedures involved in the design of experiments and be able to communicate with both statistician and experimenter in order to critically analyze the test results and to make an accurate decision.

Example 3-2 offers an opportunity to discuss some problems associated with integrating reliability into design by testing. First, the above test of shrink-fit joints was not a test for reliability. It was implied that if it would be possible to increase the coefficient of friction, then a shorter joint could be designed for a given external load or, alternatively, the static strength of the joint could be increased. If the coefficient of friction is increased, the designer has more confidence in the joint because the level of reliability has increased. However, absolute level of reliability remains unknown. The reliability of fit in the above test could be found if all specimens were subjected to a simulated external load (note that not only the torque but also the bending and tensile forces affect the shrink-fit reliability) and corresponding failure times were recorded. Two constraints are always in the way of reliability testing: time and cost. To get statistically meaningful experimental results, the experimenter is bound to test a specific number of items. If the effect of more than one factor is to be investigated, the number of items to be tested becomes greater. Usually sequential tests are performed because simultaneous tests would require identical testing equipment. It becomes clear that full-scale reliability tests are time prohibitive and thus may freeze the design process. The costs are comprised of those for the tests themselves plus losses due to possible interruption of the design process. So, although it is the statistician who makes sure that the methodology of testing is adequate, and the experimenter who makes sure that the test itself is accurate, it is the design engineer who has to decide if and what kind of test is needed.

Another problem raised by the above example is the validity of test results. The point is that the larger the sample size, the more distinctive features of the distribution can be found. For example, when comparing statistics of two distributions (say, means), the hypothesis that in two distributions the means are equal is found to be not true, whereas the difference may be so small that it should be disregarded. Similar phenomena may take place, as in the example above, when the variances are very small and the hypothesis becomes sensitive to a small variation in the means. In this respect, a distinction between what is "theoretically" different and what is "practically" different is required. Again, it is the design engineers who draw the line based on their personal judgment and experience.

The question of the validity of test results is closely associated with the problem of sample size. Intuitively, it is felt that the larger the sample, the more confidence can be placed in the test results. However, an increase in sample size leads to an increase in time and costs of the test. In statistics, sample size and probability of making an error in detecting a statistic with a specified accuracy and for a specified level of confidence are related. This relationship expressed graphically is called an operating characteristic curve. Although operating characteristic curves play an important role in the design of experiments, the proper presentation of this topic is outside the scope of this introductory text. What is important here is that the validity of the test results is associated with the level of confidence, which is chosen by the design engineer, and the latter decision leads to a specific sample size.

The next question, associated with the above example, is the applicability of test results to the operating conditions. In other words, the question is whether laboratory tests completely simulate operating conditions that a component is designed to withstand. It is clear that this cannot be achieved in general, since such factors as internal and external environment, interaction between components, and sometimes, operating regimes either cannot be simulated or are very expensive to simulate. This means that laboratory tests have a limited objective, namely, to increase the relative level of reliability by decreasing the negative effects of specified factors (friction, stress concentration, rate of corrosion, etc.) for a specified range of test conditions. Thus the effectiveness of a test program in terms of the applicability of its results depends on how critical the specified factors are in affecting the component reliability, and whether the range of test conditions covers the range of operating conditions. Again the choice of both critical factors and test conditions is done by the design engineer.

The applicability of test results can be greatly extended if tests are done in conjunction with theoretical analysis. For example, if a mathematical model of reliability of a shrink-fit is developed, in which the coefficient of friction (or the distribution function for this coefficient) is the only missing information, then tests for friction fill this information gap. The benefit of analytical-experimental integration in reliability is difficult to overestimate.

Another point, associated with the above example, is the worth of conducting nonstatistically designed tests. Simply speaking, would it make any sense to test only two shrink-fits in the above example, one with and another without a copper film? The answer is positive, provided that some conditions are met: (1) the two shrink-fits must belong to the same block, i.e., a single pin is machined and a single bushing is machined and then they are both cut in half, (2) the diameters of the pin and hole are accurately measured, and (3) the mathematical model of the dismantling force is expressed as a function of the coefficient of friction. Then by measuring the dismantling force, a coefficient of friction for each fit can be found. The effect of the copper film is thus isolated and assessed.

The scatter of the friction coefficients in the above example may not be due to the scatter in the film thickness but rather to the scatter of the material interference. However, the latter is clearly determined by the tolerances on pin and hole diameters, and their effect can easily be expressed in statistical measures. If this is the case, then the need for a statistically designed experiment disappears.

If properly formulated, the nonstatistical test can be a valid tool in the reliability program. Thus it is necessary to identify exactly what information is sought and to specify the means (analytical, experimental) to find it. The optimum solution is when the means and the goals are in agreement.

The final point that should be made with respect to the above example is the effect of the number of factors on the test design. Let us assume that the design engineer wanted to vary two factors: film thickness and surface roughness. If these two were statistically independent, then the problem could be reduced to two independent one-factor tests. However, if statistical independency cannot be postulated, then a two-factor statistical test should be done. As a result of such tests, the degree of statistical dependency can be found. This is important, since it will affect specifications of the surface roughness and film thickness, giving the highest coefficient of friction. The procedures for the multifactorial tests are well developed; however, their discussion is outside the scope of this book.

3-4 ALTERNATIVE TESTING TECHNIQUES

The need to test a system comes from the decisions made about the components and their arrangement into an assembly. It is much more difficult for design engineers to develop a "feeling" of confidence about the assembly even if they have confidence about the components in this assembly. In complex machines with hundreds of parts, it is simply impossible. It seems then that more emphasis should be placed on the testing of systems or subsystems. This need, as mentioned above, comes into conflict with time and cost limitations.

> **Example 3-3** A design engineer, in order to meet the size and weight constraints, decides to eliminate the inner ring in a ball bearing and to use the shaft as an inner ring instead (see Fig. 3–6). Although the idea seems attractive, the reliability of the support is in question, since there is no past experience with this type of design. There is a definite need to test this design for reliability. The estimation shows that at 1000 rpm the support may last for 5000 hours, which is slightly more than 200 days of testing. Let us assume that the objective of the tests is to find the best technological method of ball race

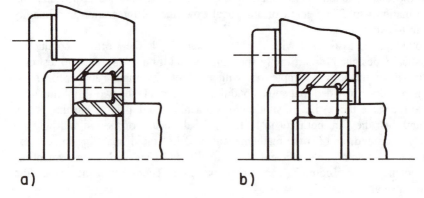

a)   b)

Figure 3-6 Two alternative designs of a shaft support.

hardening (say, gas carburizing or induction hardening are considered) for a given shaft material and maximum load.

To conduct a statistically designed test with a given degree of confidence, the number of shafts required may exceed 20, which translates into years of testing. It is clear that this approach is impractical. Does it mean that the design engineer should reject the idea? A positive answer means stopping the progress, whereas a negative answer means taking a risk. A risk is often taken, but it should be a calculated risk.

The above example illustrates the essence of the problem: the more innovative and complicated the system, the more need to test it and the more time and investment required. These conflicting tendencies are at the core of reliability problems.

Accelerated Testing

One approach to overcoming the lengthy time of testing is to accelerate it. The idea is to increase the load, speed, pressure, etc., so that the degradation processes (fatigue, wear, etc.) go faster, and thus failure occurs earlier. This is called accelerated testing. The problem with accelerated testing is how to relate the time of failure in the accelerated test to that under normal operating conditions. If a phenomenon of failure can be described by a similarity law, i.e., by a number of nondimensional parameters making up a unique unit that remains constant for both accelerated and normal test situations, then a relationship between the two failure times is easily established. Let us clarify it with a simple example.

Example 3-4 Let us consider a phenomenon of wear in a brake (Fig. 3-7) under the conditions of constant braking force P and constant speed of rotation ω. Assume that the rate of wear is proportional to the force and speed of rotation.

$$\frac{dW}{d\tau} = aP\omega \qquad (3\text{-}19)$$

where a is the coefficient of proportionality. Then the wear itself is

$$W = aP\omega\tau \qquad (3\text{-}20)$$

A specific amount of wear in a pad $W = W_{max}$ constitutes brake failure. Since coefficient a in Eq. (3-19) is supposed to be unknown, the time to failure τ_f cannot be found from Eq. (3-20) when $W = W_{max}$. If under normal operating conditions, $P = P_0$ and $\omega = \omega_0$, the time to failure $\tau = \tau_{f0}$ is long enough,

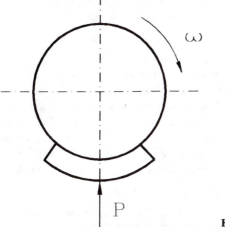

Figure 3-7 A shoe brake.

then an accelerated test can be performed under more severe load and speed conditions, $P = P_a > P_0$ and $\omega = \omega_a > \omega_0$, and the time to failure $\tau = \tau_{fa}$ can be found. Since the ratio of critical wear,

$$\frac{W_0}{W_a} = \frac{a_0 P_0 \omega_0 \tau_0}{a_a P_a \omega_a \tau_a} = 1 \tag{3-21}$$

must be the same, since $W_0 = W_a = W_{max}$ and assuming that coefficient $a = a_0 = a_a$ remains the same, the ratio between the two times to failure is found:

$$\frac{\tau_0}{\tau_a} = \frac{P_a \omega_a}{P_0 \omega_0} \tag{3-22}$$

By increasing P_a and ω_a (or one of these parameters), the time to failure τ_a in an accelerated test can be made smaller.

Note that in this case if the assumption of a in Eq. (3-19) being a constant is valid, then there is no need for an accelerated test. A (very short) test determining only the rate of wear is needed. Then from Eq. (3-19) a constant is determined, and from Eq. (3-20) a time to failure for a given $W = W_{max}$ is found. However, if the assumption of a constant rate of wear is not valid, then the relationship Eq. (3-22) is not valid either, i.e., the whole idea of the accelerated test cannot be applied.

In most practical situations, degradation processes are affected by many factors, and no explicit similarity criteria can be formulated. The failure times in accelerated and normal test conditions cannot be related exactly. If, however, a qualitatively correct mathematical model of the phenomenon is known, then some parameters of such a model may be found by conducting tests under more severe loading conditions, provided that the model remains valid for these conditions. If, for example, it is known that the fatigue crack growth rate is a function of the stress state around the crack tip, then if a maximum crack length constitutes a failure, a form of a relationship between the number of cycles to failure and the stress level can be established. The latter relationship may contain constants that are specific for a given material, size, surface finish, etc. These constants (or their distribution) can be found by conducting high-stress (accelerated) tests, and then, if distribution functions for constants can be assumed to be independent of the stress level, they can be used to calculate the statistics of the number of cycles to failure under normal stress conditions. However, if in this example, another factor, say, temperature, affects the rate of crack growth differently at different stress levels, then the above approach cannot be used.

The method of accelerated testing thus offers an opportunity to save time and costs of experiments. However, it should be done with a clear understanding of the phenomenon, the parameters affecting failure, and the existence of similarity.

Bayesian Approach

Another approach to overcoming the time and costs constraints of reliability testing is to perform a small number of tests (one or two) and then make a decision about the reliability of the entire population. An approach that formally combines objective and subjective information in order to predict the probability of failure, and also updates this estimation of probability if new information becomes available, is based on the Bayesian formula of conditional probability. Here, only an illustration of the Bayesian approach in reliability is given by a simple example for a discrete probability distribution.

Example 3-5 Consider again a shaft on two supports in Fig. 3-6. A design engineer knows that a standard ball bearing for a given load and speed of rotation may be expected to operate for 5000 hours

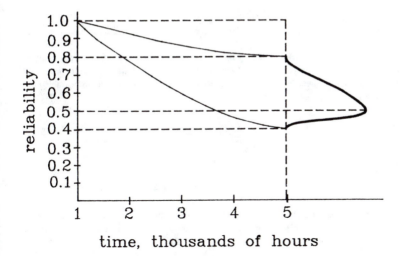

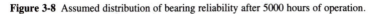

Figure 3-8 Assumed distribution of bearing reliability after 5000 hours of operation.

with the probability 0.8. However, if the shaft is used as an inner ring, then to achieve 5000 hours of operation will be less probable. In the engineer's judgment the effect of the increased contact stresses and inadequate quality control of the surface hardness should decrease the chances of successful operation during 5000 hours.

The question is, what would the level of reliability for a new design at $t = 5000$ hours be? A design engineer assumes that it may range from 0.4 to 0.8 with, most likely, a level of 0.5 (see Fig. 3-8). The probabilities of having that or another level of reliability at 5000 hours are chosen subjectively based on the designer's experience and knowledge of the design. The only requirement is that there is a 100 percent chance that the true reliability will be within the 0.4–0.8 interval specified by the engineer. Instead of a continuous distribution assumed in Fig. 3-8, let us consider a corresponding discrete distribution (see Fig. 3-9) satisfying the following requirement,

$$\sum_{i=1}^{n} P(R_i) = 1 \tag{3-23}$$

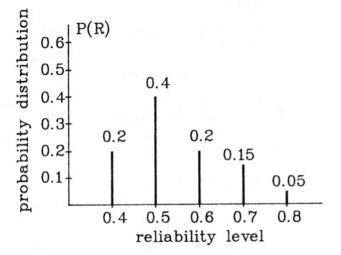

Figure 3-9 Prior probability distribution of bearing reliability after 5000 hours of operation.

where n is the number of discrete points in which the 0.4–0.8 interval is divided. Let us define two events: R_i is the event that reliability level R_i has been achieved, and E is the experimental outcome, which is success or failure.

Assume that it is of interest to know the change in the mean value of the guessed reliability if one, two, or more experiments are conducted. The information available prior to the experiments is that assumed by the design engineer (see Fig. 3-8). The mean value of a discrete random variable R_i at 5000 hours is given by

$$\mu_R = \sum_{i=1}^{4} R_i P(R_i) \tag{3-24}$$

or, using data in Fig. 3-8,

$$\mu_R = (0.4)(0.2) + (0.5)(0.4) + (0.6)(0.2) + (0.7)(0.15) + (0.8)(0.05) = 0.545$$

The question is how the value of μ_R based on the subjective information will be affected by the objective information that becomes available from a single test. The answer is found by using Bayes formula of conditional probability. In our case for a discrete random variable, Bayes formula may be written

$$P(R_i|E_i) = \frac{[P(R_i)P(E_i|R_i)]}{\left(\sum_{j=1}^{n} P(R_j)P(E_i|R_j)\right)} \qquad i = 1, 2, \ldots, n \tag{3-25}$$

$P(R_i|E)$ is the posterior (after the test) probability of $R = R_i$ given the experimental result E (success or failure)

$P(E|R_j)$ is the conditional probability of the experimental outcome E given that $R = R_j$

$P(R_j)$ is the prior (before the test) probability of $R = R_j$.

In our case the examples of prior probabilities are (let us denote by E^s a successful event and by E^f a failure to operate for 5000 hours):

1. $P(E^s|R_1 = 0.4) = 0.4$, which means that the probability of success at the reliability level $R_1 = 0.4$ is the reliability itself according to a judgment by the design engineer.
2. $P(E^f|R = 0.4) = 0.6$, which means that the probability of failure at the reliability level $R_1 = 0.4$ is $F_1 = 1 - R_1 = 0.6$.
3. $P(R = 0.6) = 0.4$, which means that the probability of having the reliability of the bearing $R_2 = 0.5$ after 5000 hours is 0.4 according to a judgment by the design engineer.

Let us assume that a single test was conducted and it was successful. Formula Eq. (3-25) can be used to reestimate the prior probabilities shown in Fig. 3-9:

$$P(R_1 = 0.4|E^s) = P(R_1 = 0.4)P(E^s|R_1 = 0.4)\bigg/ \sum_{j=1}^{5} P(R = R_j)P(E^s|R = R_j)$$

$$= \frac{(0.2)(0.4)}{(0.2)(0.4) + (0.4)(0.5) + (0.2)(0.6) + (0.15)(0.7) + (0.05)(0.8)}$$

$$= 0.0816$$

Similarly,

$$P(R_2 = 0.5|E^s) = 0.367$$

$$P(R_3 = 0.6|E^s) = 0.244$$

$$P(R_4 = 0.7|E^s) = 0.193$$

$$P(R_5 = 1.0|E^s) = 0.073$$

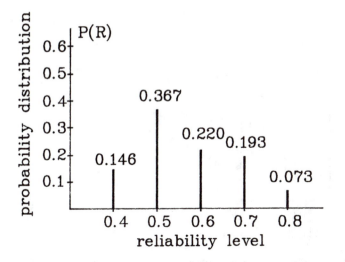

Figure 3-10 Posterior probability distribution of bearing reliability after one successful test.

The new distribution of probabilities of achieving a specified level of reliability is shown in Fig. 3-10. A calculation of the corresponding mean value gives $\mu_R = 0.567$. It is seen that a single successful test increases the level of confidence in the design: the mean value has increased from $\mu_R = 0.545$ to $\mu_R = 0.567$.

If a second test is conducted and is also successful, then the new probabilities are

$$P(R_1 = 0.4|E^s) = (0.146)(0.4)/[(0.146)(0.4) + (0.367)(0.5) + (0.22)(0.6) + (0.193)(0.7)$$
$$+ (0.079)(0.8)] = 0.044$$

$$P(R_2 = 0.5|E^s) = 0.323$$

$$P(R_3 = 0.6|E^s) = 0.232$$

$$P(R_4 = 0.7|E^s) = 0.238$$

$$P(R_5 = 0.8|E^s) = 0.103$$

The corresponding mean value has increased again to $\mu_R = 0.591$. The distribution of probabilities after the second successful test is shown in Fig. 3-11.

Let us assume, however, that the first test was not successful. Then in Eq. (3-25), $E = E^f$ and the corresponding probabilities are

$$P(R_1 = 0.4|E^f) = P(R_1 = 0.4)P(E^f|R = 0.4) \bigg/ \sum_{j=1}^{5} P(R = R_j)P(E_1|R = R_j)$$
$$= (0.2)(1 - 0.4)/[(0.2)(1 - 0.4) + (0.4)(1 - 0.5) + (0.2)(1 - 0.6)$$
$$+ (0.15)(1 - 0.7) + (0.05)(1 - 0.8)] = 0.264$$

$$P(R_2 = 0.5|E^f) = 0.440$$

$$P(R_3 = 0.6|E^f) = 0.176$$

$$P(R_4 = 0.7|E^f) = 0.099$$

$$P(R_5 = 0.8|E^f) = 0.022$$

The distribution of probabilities is shown in Fig. 3-12. The mean value of reliability becomes $\mu_R = 0.455$, which is less optimistic than the $\mu_R = 0.545$ assumed by the design engineer.

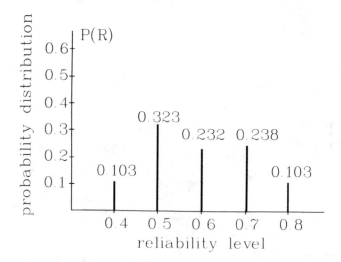

Figure 3-11 Posterior probability distribution of bearing reliability after two successful tests.

A combination of subjective and objective information in the estimation of reliability has its merits given the lack of information and difficulty of obtaining it. However, it should be used with caution, remembering the subjective part of the input data. At the same time, the more experiments performed, the more objective the final distribution of probability becomes. This transformation of subjective into objective information is a function of the number of tests and the initial guessed distribution.

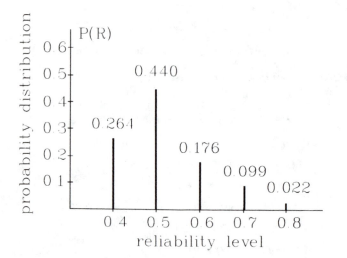

Figure 3-12 Posterior probability distribution of bearing reliability after one unsuccessful test.

PROBLEMS

3-1 Identify what is involved in the development, qualification, and acceptance tests of a washing machine. What would a reliability test involve?

3-2 Consider a washing machine. What would be the difference between the two reliability test methodologies: (*a*) to take a sample of washing machines and life test them under various load and temperature conditions, and (*b*) to conduct statistically designed tests for various load and temperature conditions. What are the advantages and disadvantages of the two approaches?

3-3 Two depths of carburization are proposed for the gear teeth. Explain how to design a statistical experiment to verify the effect of carburization depth on the gear reliability so that the concepts of randomization and blocking are implemented.

3-4 Reliability tests of two sets of gears having two different depths of carburization, h_1 and h_2, $h_1 < h_2$, were conducted. For the 10 pairs the results of the surface durability tests are shown in Table P3-4.

Table P3-4 Megacycles to Failure

					Block					
	1	2	3	4	5	6	7	8	9	10
h_1	21	23	23.5	24.5	24.6	24.7	25	27.8	28	30
h_2	23	24.6	25	27	27.1	27.2	27.7	28	30.1	31

Check the test hypotheses:

$$H_o: \mu_{h1} = 0.9\mu_{h2}$$

$$H_A: \mu_{h1} \neq 0.9\mu_{h2}$$

Here μ_{h1} and μ_{h2} are the average numbers of megacycles to failure for two depths of carburization h_1 and h_2, respectively. Assume a 95 percent confidence level.

3-5 For the data given in Problem 3-4, find a coefficient $\alpha < 1$ in the null hypothesis $H_o: \mu_{h1} = \alpha \mu_{h2}$ such that the null hypothesis is rejected. Assume a 95 percent confidence level.

3-6 Specify the conditions for accelerated testing of a household washing machine.

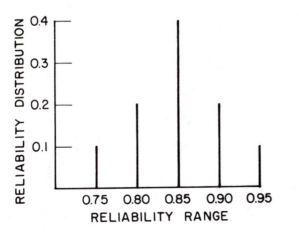

Figure P3-8

3-7 What factors affect the relationship between the time to failure of a ball bearing in normal operating conditions (constant load and speed) and in accelerated test conditions (load remains the same, but the speed is higher)? Discuss the validity of conducting an accelerated test.

3-8 A design engineer expects that a universal joint if used in more load-intensive regimes compared with its previous operation has a level of reliability, after 10^5 cycles, between 0.95 and 0.75. The engineer's assumptions on the distribution of probabilities of various levels of reliability are shown in Fig. P3-8. How would this distribution change if in four successive tests the number of cycles has exceeded the 10^5 level?

FOUR

OPERATIONAL RELIABILITY

4-1 INTRODUCTION

A design engineer has a natural interest in the performance of his or her design. There are a few reasons for such an interest. First, any failure that falls short of the expected time of operation indicates that there is a flaw somewhere in the design-manufacturing-service process. The immediate objective then is to identify the source of the flaw, which is equivalent to identifying the reason (or reasons) for the failure. Second, the design engineer is interested not only in failures as such but also in the distribution of failures in time, which is important for planning maintenance. Finally, any new design supplemented by information on its performance in time widens the designer's experience and knowledge and, as a result, allows similar products to be designed faster and more reliably.

An established feedback mechanism providing a design engineer with the information about product performance is thus an essential element of product improvement and efficient utilization. The feedback mechanism includes data collection, processing, and transfer. It also includes what data should be collected and how it should be processed. The output of this process becomes input for a design engineer when it is time to decide what kind of changes to make. In this respect it is very important for a design engineer to understand what is involved in the reliability analysis of operating systems.

4-2 FAILURE CAUSES AND TRENDS

From the designer's point of view, all failures can be divided into two categories: expected and unexpected. The expected failures are those that are in agreement with the predicted lifetime. The unexpected failures are those that fall short of the designer's prediction. Naturally, the concern is with the unexpected failures and their cause.

The unexpected failure may be caused by an error in any element of the design-manufacturing-service chain. Tracing back the source is not always possible due to disintegration of the product, extensive damage of many interacting components, difficulty of identifying the triggering mechanism, etc. Let us consider an illustrative example of various causes of failure.

Example 4-1 Consider a simple paddle-wheel fan (Fig. 4-1) that rotates with high speed and operates in a corrosive environment. The blades, gussets, and hub are welded together. Wheel failure may be caused by faulty design, manufacturing, or improper application of the fan in service.

The fan in Fig. 4-1 is a complex, three-dimensional structure mounted on a shaft. A design engineer understands that a blade may fail either from corrosion or fatigue. By specifying a material for a given operational environment, the design engineer estimates the lifetime of the blade in the presence of corrosion. Fatigue is caused by the oscillations of the blade as part of the blade-wheel-shaft-bearings-pedestal system. The level and the mode of vibration depend on the mechanical imbalances in a rotating system and on the interaction of the blade with the gas flow. The blade is subjected to high steady state stresses caused by centrifugal force and to superimposed oscillating stresses. The latter alternate with a frequency equal to or higher than the frequency of shaft rotation. The uncertainty in fatigue prediction is associated with the uncertainty of the level of alternating stresses combined with the uncertainty of the fatigue properties of welding (assuming that the fatigue properties of the blade material are better understood).

The design engineer decides that his first objective should be to avoid resonance in blades caused by rotor imbalance. Structural analysis of the fan allows the geometry of the blades and gussets to be chosen such that the fan operating speed does not cause blade resonance. However, there are other causes of oscillation, basically, blade-gas interaction and gas pulsation. Frequencies of these oscillations are higher than the frequency of shaft rotation. Since the phenomenon of blade-gas interaction was not well understood, the design engineer did not estimate the effect of the corresponding stresses on blade fatigue nor take into account the effect of the mutual corrosion-fatigue influence on blade lifetime. With all these uncertainties left, the fans were manufactured and marketed with a 1-year warranty.

Reclamations started after the first month of fan operation. From the reports, the design and reliability engineers found that most of the failures were fatigue-type; furthermore, some were low-cycle failures, while others were high-cycle. All failures took place either near the blade-hub welding or at the welding. After analyzing the operational conditions of all failed fans, the engineers found that most low-cycle fatigue failures occurred in fans operating in excessively dirty environments.

An analysis of welds by a materials science engineer revealed that some welds had failed because of weld defects. Others that failed without noticeable signs of defects showed extensive corrosion. A further check made it clear that the welding materials were not corrosion resistant in a specific appli-

gusset

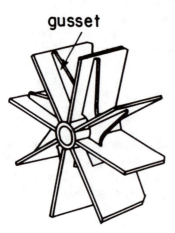

Figure 4-1 A paddle-wheel fan with one gusset.

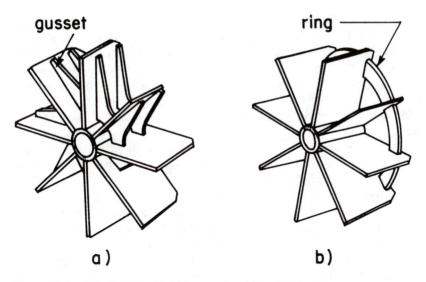

gusset

ring

a)

b)

Figure 4-2 A paddle-wheel fan with (*a*) two gussets and (*b*) a stiffening ring.

cation, and as a result, the low-cycle failure took place. The high-cycle failures in and near the weld-ings were suspected to be the result of a high-frequency resonance.

All failures were divided in three categories by cause: (1) improper application in service (dirty environment), (2) errors in manufacturing (improper welding materials for a corrosive environment), and (3) supposedly faulty design (natural frequencies close to operating speeds).

As a result of failure analysis, the design, reliability, and materials science engineers made the following recommendations: (1) to limit the area of fan applications until a modified design capable of operating in a dirty environment is introduced, (2) to change the welding materials and test the weld in various corrosive environments, and (3) to measure the dynamic properties of blades under operating conditions. All these recommendations were implemented. The dynamic tests showed that indeed the blades were resonating. The design engineer decided to test two alternative solutions: a wheel with one more gusset (Fig. 4-2*a*), and a wheel with a stiffing ring (see Fig. 4-2*b*). Additional tests proved the solution with two gussets gave better results, and it was implemented as a design modification.

In Example 4-1, all failures caused by inappropriate welding materials or appli-cations in inappropriate conditions were unexpected failures. Failures caused by in-adequate design were also unexpected; however, they were more difficult to trace and correct. In general, unexpected failures are caused by either errors in specifications in the design-manufacturing-service process or violations of otherwise correct specifica-tions. Some severe errors, like using improper welding materials, will show up early in product life, whereas others, like inadequate annealing temperatures in a process of eliminating residual stresses in a welding, may show up much later. The severe errors are usually corrected because they are systematic, i.e., errors that will be repeated in all products, whereas errors caused by failure in the quality control system or in the operating conditions for a specific product cannot be corrected, since these are not systematic.

If the systematic errors are corrected, the failure rate at the beginning of the product life will decrease. However, later in product life the nonsystematic errors will start showing up. Since these failures are caused by a multiplicity of factors, each affecting product reliability differently, they occur at random times. The failure rate is more or less constant. Thus the two types of errors in product design, manufacturing, or service—systematic and nonsystematic—cause two distinct failure rate trends.

At the later stages of product life, components come to an end of their useful life, and expected failures start taking place. It can be one or many components whose expected useful life is close enough that failures in a population of products will fall within a relatively short time interval, thus causing an increase in the failure rate. This increase indicates the beginning of the end of useful life and is another distinct failure rate trend.

Failures corresponding to three distinct failure rate trends are called early, random, and wear-out failures. The failure rate trends are depicted schematically in Fig. 4-3, where the corresponding failure types are indicated.

The first part of the operating life is characterized by "debugging," i.e, when deficiencies caused by systematic errors are discovered and corrected. Until then, full-scale production is not economical. The third part of the operating life is when too many failures make further utilization of the product uneconomical. Thus only the middle part of the product life, which is associated with the constant or near-constant failure rate trend, constitutes the useful life of the product.

The characteristic pattern of failure rates depicted in Fig. 4-3 for a product without maintenance remains qualitatively the same for the same product with maintenance if failed components are replaced and the product is returned into service. Let us assume that there is one "weak" component with a relatively short lifetime compared with that of other components and interfaces. The beginning of the wear-out phase of the product life will coincide with that for a weak component. Let us assume that wear-out starts at time t_2 (see Fig. 4-4). If at this time, maintenance is provided, i.e., all weak components of this type are replaced in all products; then the failure rate for a given population will first decrease due to the maintenance and then will converge to a constant rate until the next weak component comes to a state of wear-out (t_3 in Fig. 4-4). Thus by providing maintenance, the useful life was extended from $t_2 - t_1$ to $t_3 - t_1$. Of course, this is an idealized situation.

In reality, in a system with many components the beginnings of wear-out phases for individual components will be random, and ideal maintenance is impossible. The objective of the design for reliability in this respect is to try to break down the entire system into subsystems of components in such a way that within each subsystem the

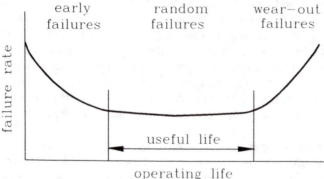

Figure 4-3 Various failure rate trends during the product life.

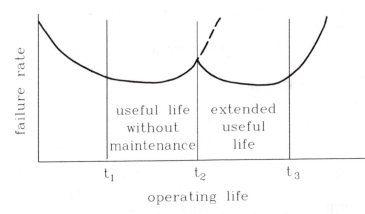

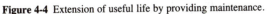

Figure 4-4 Extension of useful life by providing maintenance.

lifetimes of all components are approximately the same. For each group, maintenance is provided at the same time. If it was possible to design different components with the same reliability functions, then it would be possible to have the most economical maintenance and product utilization situation. Otherwise, components will either be allowed to fail or will be replaced when they are still in good operating condition.

4-3 SOME MATHEMATICAL MODELS OF FAILURE RATES

Statistical analysis of failure rates was discussed in Chapter 1, where a relationship between the failure rate, failure density function, and reliability function was derived [see Eq. (1-12)]. However, statistical data presented in the form of graphs cannot be used for the analysis of similar systems unless analytical models approximating experimental results are found. An analytical model is called a probabilistic model of the physical variability and involves an appropriate probability distribution function.

The problem of finding a mathematical model for a given set of data can be reduced to a problem of matching a known model against these data. There are a number of mathematical models flexible enough to describe most experimental data. However, here only the three most widely used continuous distribution models will be considered: exponential, normal, and Weibull. The estimation and the meaning of model parameters will be discussed, as well as the application of the above models to reliability problems.

Exponential Distribution

The distribution is called exponential when the time to failure is described by the exponential failure density function,

$$f(t) = \frac{1}{\lambda} \exp\left(-\frac{t}{\lambda}\right) \qquad t \leqq 0 \qquad \lambda > 0 \tag{4-1}$$

where λ is the parameter of distribution.

The reliability function is found by substituting Eq. (4-1) into Eq. (1-6):

$$R(t) = \int_t^\infty \frac{1}{\lambda} \exp\left(-\frac{\tau}{\lambda}\right) d\tau = \exp\left(-\frac{t}{\lambda}\right) \tag{4-2}$$

The hazard function is given by Eq. (1-12):

$$h(t) = \frac{f(t)}{R(t)} = \frac{1}{\lambda} \tag{4-3}$$

and is constant.

The MTTF is found from Eq. (1-23):

$$\text{MTTF} = \int_0^\infty R(\tau)\, d\tau = \int_0^\infty \exp\left(-\frac{\tau}{\lambda}\right) d\tau = \lambda \tag{4-4}$$

The plots of $f(t)$, $R(t)$, and $h(t)$ are shown in Fig. 4-5. It is seen that the failure rate in the case of the exponential distribution remains constant. Since the constant failure rate is a characteristic of the useful life of the product, the exponential distribution is thus the most useful distribution in application. At the same time it is the most simple to deal with.

In applications the mean time between failures (MTBF) is an important statistical measure. The notions of the MTTF of a component, MTTF of a system, and the MTBF are worth clarification. In this respect, let us consider a qualitative example before introducing quantitative relationships for the case of random failures.

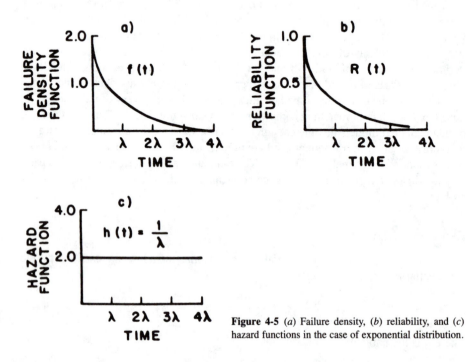

Figure 4-5 (a) Failure density, (b) reliability, and (c) hazard functions in the case of exponential distribution.

Example 4-2 In a fleet of cars, all failures during a specified number of kilometers were recorded, specifically, the type of failed component (oil filters, plugs, brakes, shock absorbers, etc.), the time (in kilometers) to failure, and the generation of the component, i.e., original or second generation if the original has failed. If all this information is available, then for each component a distribution of failure times can be extracted from the records (note that for a second, third, etc., generation the time starts from the moment of installation). When this information is statistically processed, the MTTF for each component can be found [see Eq. (1-25)]. In the case of a constant failure rate the average of all failure times for a component is the estimation of the MTTF. So the MTTF in the above sense is the mean time to failure from the moment of installation of the population of identical components in identical conditions, irrespective of the absolute time of the component failure in the system.

At the same time, a failure of the car can be viewed as an event by itself without specifying what caused this failure. Then from the data above, information can be extracted about the times to the first failure of all cars, second failure (after repair), and so on. Also the averages of times to the first failure, second failure, etc., can be found. Then the average of times to the first failure is the mean time to failure of the system (MTTF$_s$), while the average of all averages (to the first failure of all cars, to the second failure of all cars, etc.) is the mean time between failures (MTBF).

In a complex system the MTBF is used for maintenance planning. There is a simple relationship between the MTTF of individual components and MTTF$_s$ for a system made up of these components in the case when all components are in series and in a state of random failure. The system reliability is given in this case by Eq. (2-6):

$$R_s(t) = \prod_{i=1}^{n} R_i(t) \tag{4-5}$$

Since all components are supposed to be in a state of random failure, each $R_i(t)$ in Eq. (4-5) is described by the exponential function Eq. (4-2) with an individual parameter λ_i. Equation (4-5) then becomes

$$R_s(t) = \prod_{i=1}^{n} \exp\left(-\frac{t}{\lambda_i}\right) = \exp\left(-t \sum_{i=1}^{n} \frac{1}{\lambda_i}\right) \tag{4-6}$$

which means that the reliability of the system is also described by the exponential function with the parameter

$$\frac{1}{\lambda_s} = \sum_{i=1}^{n} \frac{1}{\lambda_i} \tag{4-7}$$

The parameter λ_s represents the MTTF$_s$. So Eq. (4-7) can be rewritten as

$$\frac{1}{\text{MTTF}_s} = \sum_{i=1}^{n} \frac{1}{\text{MTTF}_i} \tag{4-8}$$

Formally, Eq. (4-7) gives the impression that as long as λ_s is found, then a system can be considered as an equivalent component described by the exponential distribution function. It is interesting that MTTF$_s$ < MTTF$_{min}$, where MTTF$_{min}$ characterizes a component with the minimum λ. The inequality means that an overlapping of failures of otherwise reliable components results in a shorter average time for system failure. This is equivalent to the effect of multiplication of reliabilities in a series functional diagram.

Let us consider now the reliability of the system in a random state of failure after failed components in some products have been replaced. The system reliability is then given by

$$R_s(t) = \prod_{i=1}^{n} R_i(t - t_i) \qquad t \geq t_i \tag{4-9}$$

where t_i is the time to the first failure of the ith component. Note that $t_i = 0$ if the component belongs to the first generation.

The reliability of each component is now written as

$$R_i(t - t_i) = \exp\left(-\frac{t - t_i}{\lambda_i}\right) \qquad t \geq t_i \tag{4-10}$$

A substitution of Eq. (4-10) into Eq. (4-9) gives

$$R_s(t) = \prod_{i=1}^{n} \exp\left(-\frac{t - t_i}{\lambda_i}\right) = \exp\left(-t\sum_{i=1}^{n}\frac{1}{\lambda_i}\right)\exp\left(\sum_{i=1}^{n}\frac{t_i}{\lambda_i}\right) \tag{4-11}$$

The first factor in Eq. (4-11) is, according to Eq. (4-7), $\exp(-t/\lambda_s)$, whereas the sum in the second factor can be represented in the form

$$\sum_{i=1}^{n}\frac{t_i}{\lambda_i} = \mu_{T1}\sum_{i=1}^{n}\frac{1}{\lambda_i} = \frac{\mu_{T1}}{\lambda_s} \tag{4-12}$$

where μ_{T1} is the weighted average of all times to the first failure.

A substitution of Eq. (4-12) into Eq. (4-11) represents the system reliability after first repairs in the form

$$R_s(t) = \exp\left(-\frac{t - \mu_{T1}}{\lambda_s}\right) \qquad t \geq \mu_{T1} \tag{4-13}$$

It is seen that μ_{T1} indicates a time with respect to which the system reliability after the first repairs is measured.

Similarly, if t_i in Eq. (4-9) represents the time to the second failure in an absolute time scale, then the weighted average of all times to the second failure, μ_{T2}, can be found:

$$\mu_{T2} = \lambda_s\sum_{i=1}^{n}\frac{t_i}{\lambda_i} \tag{4-14}$$

If m generations of parts were replaced, then the average of all weighted averages is the mean time between failures:

$$\text{MTBF} = \frac{1}{m}\sum_{j=1}^{m}\mu_{Tj} \tag{4-15}$$

Note, that if all t_i are equal, i.e., parts are replaced at the same time during maintenance, then $t_i = t_{ri} = $ const in Eq. (4-12), and it follows that $\mu_{T1} = t_{r1}$, i.e., the weighted average time is equal to the time of replacement. Similarly, $\mu_{T2} = t_{r2}$ in Eq. (4-14). In this case the MTBF in Eq. (4-15) simply becomes the average time between maintenance.

The confidence limits for λ are based on a specified level of significance α and the number of observations n. A statistic for estimating of the unknown mean μ_λ^* is

$$\mu_\lambda^* = \frac{1}{n} \sum_{i=1}^{n} t_i \qquad (4\text{-}16)$$

where t_i is the time to failure of the ith element out of the n in a sample. Two-sided $100(1 - \alpha)$ percent confidence limits for μ_λ are

$$\bar{\mu}_\lambda = 2n\mu_\lambda^*/\chi^2\left[\left(1 - \frac{\alpha}{2}\right), 2n\right] \qquad (4\text{-}17)$$

$$\mu_\lambda = 2n\mu_\lambda^*/\chi^2\left[\frac{\alpha}{2}, 2n\right] \qquad (4\text{-}18)$$

where $\chi^2(\delta, 2n)$ is the percentage point of the chi-square distribution with $2n$ degrees of freedom. The points are tabulated (see Appendix Table B-2).

Example 4-3 The failures of 15 shock absorbers were recorded. The results (in kilocycles) were as follows: 2.5, 3.5, 4.0, 6.5, 7.0, 9.5, 10.5, 14.0, 17.0, 19.0, 21.0, 24.0, 27.0, 32.0, and 36.0. If it is known that the shock absorbers were in a state of random failure, what would the 95 and 99 percent two-sided confidence limits on the mean life be?

SOLUTION The estimation of the mean life according to Eq. (4-16) is

$$\mu_\lambda^* = \frac{1}{15} \sum_{i=1}^{15} t_i = 15.6 \text{ kilocycles}$$

The chi-square values from Appendix B-2 are

95 percent confidence ($\alpha = 0.05$, $2n = 30$)

$$\chi^2\left[\left(1 - \frac{\alpha}{2}\right), 2n\right] = \chi^2(0.975, 30) = 16.791.$$

$$\chi^2\left[\frac{\alpha}{2}, 2n\right] = \chi^2(0.025, 30) = 46.979$$

99 percent confidence ($\alpha = 0.01$, $2n = 30$)

$$\chi^2[0.995, 30] = 13.787 \qquad \chi^2[0.005, 30] = 53.632$$

Then the 95 percent confidence limits for μ_λ are

$$10.0 \leq \mu_\lambda \leq 27.8$$

and for 99 percent are

$$8.7 \leq \mu_\lambda \leq 33.9$$

It is seen how an increase in the confidence level $(1 - \alpha)$ widens the confidence interval. A probability that the true mean is within the specified limits is higher if the range is wider.

The confidence limits for reliability are estimated based on the confidence limits for λ. The estimation of the reliability function is given by

$$R^*(t) = \exp\left(-\frac{t}{\mu_\lambda^*}\right) \qquad (4\text{-}19)$$

where μ_λ^* is taken according to Eq. (4-16).

The confidence limits for $R(t)$ are

$$\exp\left(-\frac{t}{\underline{\mu}_\lambda}\right) \leq R(t) \leq \exp\left(-\frac{t}{\bar{\mu}_\lambda}\right) \tag{4-20}$$

The 100th percentile is the age (in time, kilometers, cycles, etc.) by which a certain fraction p of the population has failed. The t_p, the time of failure, can be found, in general, from the equation

$$R(t_p) = 1 - p \tag{4-21}$$

In the case of the exponential distribution, Eq. (4-21) becomes

$$\exp\left(-t_p/\lambda\right) = 1 - p \tag{4-22}$$

from which it follows that

$$t_p = \lambda \ln\left[(1 - p)^{-1}\right] \tag{4-23}$$

and the corresponding estimation based on test results is

$$t_p^* = \mu_\lambda^* \ln\left[(1 - p)^{-1}\right] \tag{4-24}$$

The confidence limits for t_p are

$$\underline{\mu}_\lambda \ln\left[(1 - p)^{-1}\right] \leq t_p \leq \bar{\mu}_\lambda \ln\left[(1 - p)^{-1}\right] \tag{4-25}$$

Example 4-4 For the data given in Example 4-3, determine (1) a 95 percent lower confidence limit for the 10-kilocycle reliability and compare it with the estimated reliability for this number of cycles, and (2) a 95 percent two-sided confidence limit for the number of cycles at which 15 percent of the shock absorbers will have failed.

SOLUTION

(a) The estimated reliability is

$$R^*(10) = \exp\left(-10/15.6\right) = 0.526$$

Whereas the lower confidence limit is

$$\underline{R}(10) = \exp\left(-10/\underline{\mu}_\lambda\right) = \exp\left(-10/10\right) = 0.368$$

Thus there is a 95 percent chance that the true reliability after 10 kilocycles would be as low as 0.368, which is much smaller than the estimated value 0.526.

(b) The estimated number of cycles for the reliability level of 0.85 is found from Eq. (4-24):

$$t_{0.15}^* = 15.6 \ln\left[(1 - 0.15)^{-1}\right] = 2.53 \text{ kilocycles}$$

whereas the confidence limits for $t_{0.15}$ with a 95 percent confidence level are

$$1.6 \text{ kilocycles} \leq t_{0.15} \leq 4.5 \text{ kilocycles}$$

In Example 4-3 the data were supposed to be exponentially distributed. However, this assumption was never verified. In statistics the assumption is called a hypothesis, and the process of verification is called hypothesis testing (this terminology was introduced in Chapter 3). The fit of a given set of data to a specific reliability function can be found analytically with the specified degree of confidence. It can also be done using graphical methods, by which the compatibility of the model and experimental data can be assessed visually, making them simple and clear but not necessarily accurate. Here a graphical method of data assessment and parameter estimation is considered for the purpose of illustration.

Let us take the logarithm of Eq. (4-2):

$$\ln R(t) = -\frac{t}{\lambda} \tag{4-26}$$

Taking the natural logarithm again gives

$$\ln \ln \frac{1}{R(t)} = \ln t - \ln \lambda \tag{4-27}$$

On a double log-log scale, the relationship between the inverse of reliability function and the time to failure would be linear if the distribution was exponential. The parameter of the distribution can be found as a parameter of a straight line. It became customary to use a failure function, $F(t) = 1 - R(t)$, instead of a reliability function, so that Eq. (4-27) is written

$$\ln \ln [1 - F(t)]^{-1} = \ln t - \ln \lambda \tag{4-28}$$

Special probability graph paper with a double-log vertical axis and a single-log horizontal axis, in which the vertical axis is labeled in terms of $F(t)$ is called Weibull distribution probability paper (see Appendix B). Since empirical data are supposed to be used in Eq. (4-28), $F(t)$ and t are expressed as discrete functions (see Chapter 1):

$$F(t_i) = 1 - \frac{N_s(t_i) + 0.7}{N + 0.4} \tag{4-29}$$

and

$$t_i = t_{i-1} + \Delta t_i \qquad i = 1, 2, \ldots, n \tag{4-30}$$

where N is the total number of failed units during the time of observation t_n, $N_s(t_i)$ is the number of units surviving at time t_i, and Δt_i is the length of the ith time interval.

Note that t_0 is not necessarily equal to zero in Eq. (4-30). If this is the case, then Eq. (4-28) should be written

$$\ln \ln [1 - F(t)]^{-1} = \ln (t - t_0) - \ln \lambda \tag{4-31}$$

If t_0 is nonzero, then it remains unknown for a given data set. The point is that in recording failure times, a reference point is chosen arbitrary (beginning of service, for example) with respect to the physical property of the item under consideration to survive. The uncertainty associated with the reference point t_0 affects the graphical data and may distort the conclusion. If t_0 exists such that Eq. (4-31) is plotted as a straight line, then this t_0 is usually found by trial and error.

Example 4-5 Let us consider data in Example 4-3 again. Do they fit the exponential distribution, and if so, what is the value of the parameter λ? Let us take $\Delta t_i = 5$ kilocycles. Note that in this case, $n = 15$. The calculations are carried out in Table 4-1. The results are plotted on Weibull paper in Fig.

Table 4-1 Test and Reliability Data for 15 Shock Absorbers

t_i, kilocycles	$N_s(t_i)$	$F(t_i)$	$R(t_i)$	$f(t_i)$	$h(t_i)$
<5	12	0.176	0.824	—	—
5–10	9	0.370	0.630	0.0388	0.062
10–15	7	0.500	0.500	0.0260	0.052
15–20	5	0.630	0.370	0.0260	0.070
20–25	3	0.760	0.240	0.0260	0.108
25–30	2	0.825	0.175	0.0130	0.074
30–35	1	0.890	0.110	0.0130	0.118
35–40	0	0.955	0.045	0.0130	0.288

4-6 (it is assumed that $t_0 = 0$). It is seen that a straight line approximates the failure data fairly accurately, except for the last point. For the comparison $h(t)$ is calculated in Table 4-1, and the results are shown in Fig. 4-7. It is seen that $h(t)$ has a small upward trend if it was not for the last point (the out-of-sequence last point can be due to the small size of the recorded data). The simplest way to find λ is to use Eq. (4-26) for any point $t = t_i$ that lies on the straight line in Fig. 4-6. Let us take $t_i = 20$ kilocycles and the corresponding $F(t_i) = 0.63$. Then

$$\lambda = \frac{t_i}{\ln [1 - F(t_i)]} = \frac{20}{0.994} = 20.1 \text{ kilocycles}$$

This estimation gives λ larger that the mean $\mu_\lambda^* = 15.6$ kilocycles in Example 4-3. At the same time for $\mu_\lambda^* = 15.6$ the constant hazard value is $h^* = 1/\mu_\lambda^* = 0.064$ and approximates only the initial part of the failure times well (Fig. 4-7).

The discrepancy between the two values of λ estimated by two methods, graphical and numerical, may be due to the errors in plotting and interpretation of data on a graph, to the assumption that the data can be described by the exponential distribution (in fact, it cannot be), or both. There are various means of testing the hypothesis of exponential distribution that are outside the scope of this book. However, the point is that care should be taken in data interpretation.

Normal Distribution

The normal distribution of times to failure is given by

$$f(t) = \frac{1}{\sigma\sqrt{2\pi}} \exp\left[-\frac{1}{2}\left(\frac{t - \mu}{\sigma}\right)^2\right] \tag{4-32}$$

where μ is the mean time to failure and σ is the standard deviation of failure times. In this case the reliability function is

$$R(t) = \int_t^\infty f(\tau)\, d\tau = \frac{1}{\sigma\sqrt{2\pi}} \int_t^\infty \exp\left[-\frac{1}{2}\left(\frac{\tau - \mu}{\sigma}\right)^2\right] d\tau \tag{4-33}$$

A substitution of variables $z = (\tau - \mu)/\sigma$ transforms the latter integral into

$$R(z_t) = \frac{1}{\sqrt{2\pi}} \int_{z_t}^\infty \exp\left(-\frac{z^2}{2}\right) dz \tag{4-34}$$

where it is denoted

$$z_t = \frac{t - \mu}{\sigma} \tag{4-35}$$

The reliability function can be expressed through the $\phi(\zeta)$ function,

$$R(z_t) = 1 - \phi(z_t) \tag{4-36}$$

where

$$\phi(z) = \frac{1}{\sqrt{2\pi}} \int_{-\infty}^{z} \exp\left(-\frac{u^2}{2}\right) du \tag{4-37}$$

is tabulated (see Appendix Table B-1).

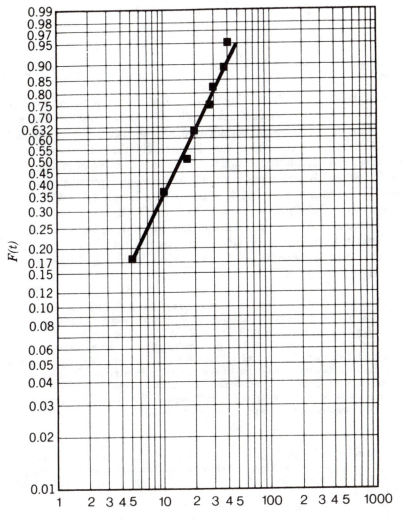

Figure 4-6 Weibull plot for shock absorbers.

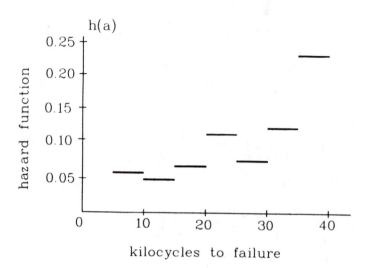

Figure 4-7 Hazard function for shock absorbers.

The hazard function is calculated using Eq. (1-12), $h(t) = f(t)/R(t)$, and it can be shown that it is an increasing function of time. In Fig. 4-8 the shapes of $f(t)$, $R(t)$, and $h(t)$ shown for three different standard deviations indicate the effect of the magnitude of the standard deviation on these functions. Theoretically, the normal distribution extends into negative time, which contradicts the physical meaning of product life, which is always positive. However, this fact can be neglected in applications if μ is at least 2–3 times greater than σ. For example, if $\mu/\sigma = 2$, then only 2.5 percent of the distribution is negative, and if $\mu/\sigma = 3$, only 0.14 percent is negative.

Because the hazard function is an increasing function of time, the normal distribution can be used to describe the wear-out portion of the product life. It can be useful in describing physical and chemical deteriorations, aging, corrosion, fatigue, creep, etc. Let us consider an example.

Example 4-6 Failures of 20 shafts operating at a constant stress level were reported. The decimal logarithm, log K, of the number of cycles to failure was distributed as follows: 4.2, 4.4, 4.58, 4.6, 4.7, 4.76, 4.81, 4.86, 4.88, 4.96, 4.98, 4.99, 5.02, 5.12, 5.17, 5.25, 5.36, 5.4, 5.49, and 5.78. (1) Take the interval $\Delta t = \Delta (\log K) = 0.2$ and plot the reliability, failure density, and hazard functions. Does it look like log K is normally distributed? (2) Assume that log K is a normally distributed variable and that the mean and standard deviations can be approximated by the average and an unbiased variance, respectively. What number of failures should be expected in a population of 100 products with similar shafts in similar conditions during the first 100 kilocycles? (3) What are the 95 percent confidence limits for the number of failures during the first 100 kilocycles?

SOLUTION

1. Test and reliability data are shown in Table 4-2. Plots of $R(t_i)$, $f(t_i)$, and $h(t_i)$ are shown in Fig. 4-9. Although the $h(t)$ function is increasing, which means that qualitatively a normal distribution for log K will not be contradictory, the $f(t)$ function for the assumed Δt does not give an indication that the normal distribution is a good fit.

2. The average value of log K is

$$\mu^*_{\log K} = \frac{1}{20} \sum_{i=1}^{20} (\log K)_i = 4.96$$

and the variance of $\log K$ is

$$S^2_{\log K} = \frac{1}{19} \sum_{i=1}^{20} (\log K - \mu^*_{\log K})^2 = 0.146$$

and according to the assumption $\sigma = S_{\log K} = 0.382$. The number of failures during 100 kilocycles is

$$N_f \text{ (number of cycles} < 10^5) = N_0 F(100) = N_0\, \phi(z_{100})$$

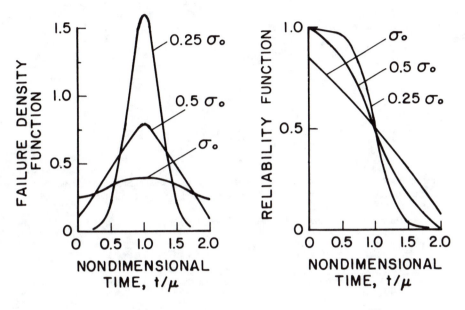

(a)

(b)

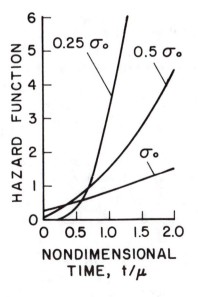

(c)

Figure 4-8 Failure density, reliability, and hazard functions in the case of normal distribution.

Table 4-2 Test and Reliability Data for 20 Shafts

$t_i = \log K$	N_s	$F(t_i)$	$R(t_i)$	$f(t_i)$	$h(t_i)$
4.2–4.4	18	0.083	0.917	—	—
4.4–4.6	16	0.181	0.819	0.49	0.60
4.6–4.8	14	0.279	0.721	0.49	0.68
4.8–5.0	8	0.574	0.426	1.48	3.47
5.0–5.2	5	0.721	0.279	0.74	2.65
5.2–5.4	2	0.868	0.132	0.74	5.60
5.4–5.6	1	0.917	0.083	0.25	3.01
5.6–5.8	0	0.966	0.034	0.25	7.35

where

$$z_{100} = \frac{\log 10^5 - \log K^*}{\sigma} = 0.105$$

From the normal table (Appendix Table B-1), $\phi(0.105) = 0.5596$, so $N_f = 56$ shafts. At the same time, if $F(t_i = 5.0) = 0.574$ is used from the test data directly, without implying a normal distribution, the number of failed shafts is 57, which is in a good agreement with the result based on a mathematical model.

3. The confidence interval on the population mean is given by

$$\mu^*_{\log K} - t_{\alpha/2, \, v,} \left(\frac{S_{\log K}}{\sqrt{20}}\right) \leq \mu_{\log K} \leq \mu^*_{\log K} + t_{\alpha/2, \, v,} \left(\frac{S_{\log K}}{\sqrt{20}}\right)$$

For $\alpha = 0.05$ and $v = n - 1 = 19$ the t value from the tables (see Appendix Table B-3) is $t_{0.025,19} = 2.093$. Then

$$4.781 \leq \mu_{\log K} \leq 5.138$$

The confidence interval on the variance is

$$\frac{(n - 1)S^2_{\log K}}{\chi^2_{\alpha/2, \, v}} \leq \sigma^2 \leq \frac{(n - 1)S^2_{\log K}}{\chi^2_{1 - \alpha/2, \, v}}$$

For $\alpha = 0.05$ and $v = n - 1$ the chi-square values are (see Appendix Table B-2) $\chi^2_{0.025,19} = 32.852$ and $\chi^2_{0.975,19} = 8.907$. Then

$$0.084 \leq \sigma^2 \leq 0.311$$

The corresponding limits for z_{100} are

$$-0.247 \leq z_{100} \leq 0.756$$

From the normal table (Appendix Table B-1), $\phi(-0.247) = 0.401$ and $\phi(0.756) = 0.7764$. Thus there is a 95 percent confidence that the number of failed shafts may be within the limits

$$40 \leq N_f \leq 78$$

This indicates the margin of uncertainty if $N_f = 56$ would be taken as a target in planning the maintenance schedules.

Weibull Distribution

In this distribution the hazard function is postulated in a form

$$\int_0^t h(\tau) \, d\tau = \left(\frac{t - \delta}{\theta - \delta}\right)^\beta \qquad t \geq \delta \geq 0 \qquad (4\text{-}38)$$

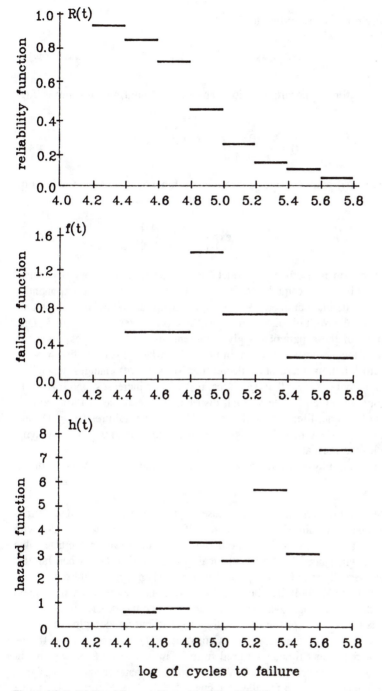

Figure 4-9 Reliability, failure, and hazard functions for shafts.

Then the reliability function according to Eq. (1-15) is

$$R(t) = \exp\left[-\left(\frac{t - \delta}{\theta - \delta)}\right)^{\beta}\right] \tag{4-39}$$

The explicit formula for the hazard function can now be found by substituting $R(t)$ into Eq. (1-13):

$$h(t) = \frac{\beta}{\theta - \delta}\left(\frac{t - \delta}{\theta - \delta}\right)^{\beta - 1} \tag{4-40}$$

And finally, having $R(t)$ and $h(t)$, the failure density function is obtained using Eq. (1-12):

$$f(t) = \frac{\beta(t - \delta)^{\beta - 1}}{(\theta - \delta)^{\beta}} \exp\left[-\left(\frac{t - \delta}{\theta - \delta}\right)^{\beta}\right] \tag{4-41}$$

The Weibull distribution is useful as a model for product life. Its popularity is explained by the flexibility of the model to fit many life data. Indeed, the exponential distribution has only the one parameter λ, the normal distribution has the two parameters μ and σ, while the Weibull distribution has three parameters, θ, β, and δ. Different combinations of these parameters give different distribution shapes, and thus properties. Some of these shapes are shown in Fig. 4-10, where it is taken that $\delta = 0$. The type of hazard function (increasing, decreasing, or constant) changes depending on the value of β, namely, if $\beta < 1$, it decreases, if $\beta > 1$, it increases, and if $\beta = 1$, it remains constant. In principle, the Weibull distribution is capable of fitting any of the three periods of product life: early failures ($\beta < 1$), random failures ($\beta = 1$), and wear-out failures ($\beta > 1$). When $\beta = 3.44$, the distribution is approximately symmetrical and is close to normal.

Let us consider the physical interpretation of the constants in the Weibull distribution.

1. The δ is the locating constant; it is expressed in the same units as t and defines the starting point or origin of the distribution. If the starting point coincides with the beginning of service, then $\delta = 0$; otherwise, $\delta > 0$. However, the true value of δ is not known, since there is no physical ground to identify δ when failures are recorded. Usually it is found by trial and error when trying to fit the experimental data into the Weibull distribution. If such δ cannot be found, then it should be assumed that the Weibull distribution cannot describe the experimental data.
2. The $\theta = \delta$ is the scaling constant, stretching the distribution along the time axis. When $t - \delta = \theta - \delta$, the reliability is constant, $R(1) = \exp(-1) = 0.368$. This constant represents time, measured from δ, by which 63.2 percent of the population can be expected to fail (63.2th percentile), whatever value is assigned to β. For this reason, $\theta - \delta$ is called the characteristic life. Figure 4-10b shows that the characteristic life relates to the point of intersection of all reliability functions.
3. The β is the shaping constant, which controls the shape of the distribution, as seen from Figs. 4-10a and 4-10c.

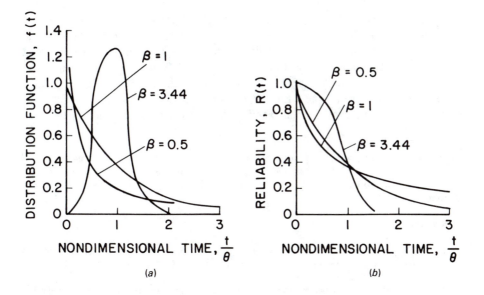

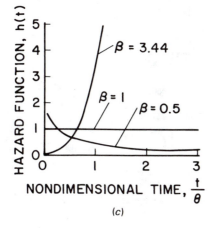

Figure 4-10 Failure density, reliability, and hazard functions in the case of Weibull distribution.

The parameters of the Weibull distribution can be found and their accuracy assessed (in a statistical sense) using analytical methods. However, herein only the graphical methods are discussed.

Let us consider Eq. (4-39) and take the double natural logarithm from both sides:

$$\ln \ln R(t)^{-1} = \beta[\ln(t - \delta) - \ln(\theta - \delta)] \qquad (4\text{-}42)$$

Usually in this equation, $R(t)$ is represented through the failure function, so that Eq. (4-42) is written in the form

$$\ln \ln [1 - F(t)]^{-1} = \beta[\ln(t - \delta) - \ln(\theta - \delta)] \qquad (4\text{-}43)$$

The latter can be viewed as a linear equation: $y = \beta x + c$, where $y = \ln \ln [1 - F(t)]^{-1}$, $x = \ln(t - \delta)$, and $c = -\beta \ln(\theta - \delta)$. On special Weibull probability paper (see Appendix C) with double-log versus single-log axes, the reliability versus time to failure for the Weibull distribution is represented by a straight line. The slope of the line determines β, while θ can be estimated by recalling that $R(t = \theta) = 0.368$, and so projecting the point $F(t = \theta) = 0.632$ on the abscissa gives the value of θ. Usually δ is assumed to be zero. If the plot is a straight line, then the assumption is correct; otherwise; the nonzero δ should be tried. If the plot still remains nonlinear, it means that the experimental data cannot be modeled by the Weibull distribution.

Example 4-7 Let us consider again the data in Example 4-5 for 20 failed shafts, and (1) make a Weibull probability plot, (2) estimate the parameters of the distribution, (3) find out the nature of the failure rate, and (4) find out whether the distribution is close to the normal, an assumption made in Example 4-5.

SOLUTION

1. The plot is constructed on the basis of data in Table 4-2 (see Fig. 4-11) assuming that $\delta = 0$. It looks like a straight line fits the experimental points, and thus the Weibull distribution can be assumed to be appropriate in this case.
2. The estimation of β can best be accomplished by taking any two points that lie on a straight line (F_1, t_1) and (F_2, t_2).

Then

$$\beta = \frac{\ln \ln (1 - F_2)^{-1} - \ln \ln (1 - F_1)^{-1}}{\ln t_2 - \ln t_1}$$

Taking from Table 4-2 $(F_1, t_1) = (0.083, 4.3)$ and $(F_2, t_2) = (0.966, 5.7)$ gives

$$\beta = \frac{1.218 - (-2.446)}{1.74 - 1.46} = 13.08$$

The θ is found from the plot: for $F(t) = 0.632$, $t = \theta = 5.0$. Here $\delta = 0$, since the line is sufficiently straight.

3. Since $\beta > 1$, the failure rate is increasing.
4. The distribution would be close to normal if $\beta \cong 3.44$. In this case it is much greater. Thus the assumption made in Example 4-5 does not seem to be satisfactory.

Note in conclusion that although the graphical assessment is advantageous as a visual aid, its accuracy might be questionable, especially for small samples.

4-4 LIFE DATA ANALYSIS VERSUS DESIGN

Some purposes of life data collection and analysis activities are the improvement of the existing product and the accumulation of knowledge for further designs. In this respect it is difficult to underestimate the need for a data collection and analysis system. However, to be effective, this system should be based on properly formulated objectives.

The statistics about product failure times can only be used for service planning. For example, a car rental company is interested in the number of cars operating at any given time. But even in this situation, failure statistics without maintenance statistics can give the answer only if maintenance does not affect the number of cars in service (i.e., a car is immediately substituted upon a failure). However, if this is not the case,

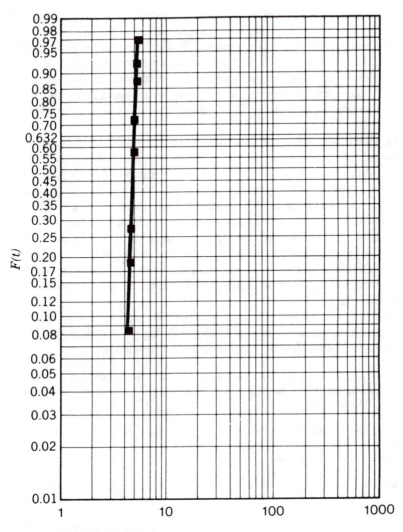

Figure 4-11 Weibull plot for shafts.

then more specific information is needed about failed parts, which affect the repair times and thus the number of cars in service.

Thus product failure times have relatively limited use by themselves. Supplemented by information about failed parts (or subsystems), the failure time statistics become more useful in two respects: (1) they allow estimation of maintenance times to be included in the serviceability analysis, and (2) they allow the failure statistics to be analyzed for parts as opposed to the entire product. The second point results in a data base concerning the reliability of specific components in specific conditions. This information can be used in further designs if parts and conditions remain the same, or even as a guideline if the conditions change.

However, information about product and part failure times does not help to improve the design efficiently unless more information about failed parts and operational conditions is available. As discussed in Chapter 2, reliability of a component depends on the properties of the component itself, on the interaction of this component with the system, and on the environmental conditions. Three different causes of component failure lead to a single act of failure. It is difficult (if not impossible) to separate the effect of each of these conditions on the fact of failure. In addition, a variety of factors can lead to the same result, such as fatigue. However, separation of factors is exactly what is needed to improve the design. The difficulty that a design engineer faces is evident.

From the point of view of design improvement the data should comprise statistical and nonstatistical information. Statistics may reveal common-cause failures due to overloading, overheating, vibrations, etc., which may indicate a need to improve either the entire system or a subsystem. Statistics may also indicate that a specific failure is a rare event, which means that an unusual cause should be looked for. Nonstatistical information concerns parts and interfaces, properties of materials really used as opposed to those that are specified, the rate or amount of wear, and the presence of cracks, corrosion, etc. Nonstatistical information is based on laboratory tests and measurements. As a result of laboratory investigation, the conformity to design specifications is checked and the number of factors responsible for component failure is reduced (or the cause identified).

Parts reliability is based on statistics, laboratory data analysis, and analysis and tests (if any) performed during the design development stage. The conclusions should also be supplemented by recommendations about corrective action.

Example 4-8 Let us consider various probable situations concerning leakage of lip-seals.

In three similar products the lip-seals developed leakage after the following times in operation: 10, 100, and 500 hours.

1. Visual inspection of the first seal, which failed after 10 hours, revealed a cut in the seal lip. Additional inspection revealed that the shaft had sharp shoulders and that the damage to the seal could have occurred during assembly.

   CONCLUSION The failure is not system-affected; however, it is not confined to a failed system, either.

   RECOMMENDATION Either use special lip-seal mounting tools or improve the shaft design by making a chamfer.

   The failure of this seal belongs to the early failure type.

2. Visual inspection of the second seal, which failed after 100 hours, did not show any signs of seal damage. However, shaft wear was noticeable. Laboratory examination showed excessive wear of the lip and also hardening of the sealing element, possibly attributed to the higher than usual internal temperature, level of vibrations, local friction, or fluid pressure. Since no other seals failed in a similar mode, the shaft-seal interface had to be examined. Measurement indicated that the shaft diameter was greater than that required by the specifications.

   CONCLUSION The failure is due to the oversized shaft diameter.

   RECOMMENDATION Improve quality control.

   The failure of the second seal also belongs to the early failure type. However, it is caused by a flaw in the manufacturing process.

3. Visual examination of the third seal, which failed after 500 hours, and the shaft did not reveal any unusual signs. Laboratory examination showed hardening of the seal element. Comparison with previously obtained seal tests at various simulated conditions showed that the degree of hardening was within the expected limits.

CONCLUSION The failure is not caused by the design or manufacturing flaws.

RECOMMENDATION Do nothing, treat this information as statistical data for future applications.

The failure of the third seal belongs to the random failure type.

It is important to note that the data about the first two failures should not become a part of the statistics concerning seal failure times because the causes of failure have been eliminated and these two seals are not a part of the population anymore.

The above example shows that the separation of failure rates into three categories (early, random, and wear-out) is only possible if the failures are examined. Since planning of maintenance is based on the random and initial stages of wear-out failures, examination of failed parts is imperative before undertaking corrective action, accumulating relevant experience, and avoiding distortion of the statistics of failures needed in planning the preventive maintenance. Thus a proper system of failure analysis is important in all respects.

PROBLEMS

4-1 In a fleet of 30 trucks, nine failures were reported after 180 days in service. If trucks are in a state of random failure, how many should be expected to be in service after 360 days? What is the MTTF?

4-2 For a boiler-piping-turbine system, shown in Fig. 2-3, it is known that the MTTF of a boiler, piping, and turbine is 3, 6, and 4 months, respectively. If it is assumed that the system is in a random failure mode, and failures of components are statistically independent, what is the reliability function of the system?

4-3 In Table P4-3, failures of dc motors in specified intervals of days are given. Assuming that the dc motors were in a state of random failure, what is the 95 percent, two-sided confidence interval for the mean life? Determine the number of days at which 10 percent of motors will have failed.

Table P4-3 Failures of dc Motors

			Interval		
	0–200	200–400	400–600	600–800	800–1000
Number of failures	5	3	2	2	1

4-4 For the data in Problem 4-3, check if the distribution is exponential by (a) using a graphical method and (b) plotting a hazard function. What is the hazard rate determined by each method?

4-5 An observation shows that a specific component has a normal distribution of failure times with $\mu = 16,000$ hours and $\delta = 4,000$ hours.

    (a) Calculate and plot the reliability and hazard functions for this distribution.

    (b) If 20 components were tested, how many would you expect to fail between 12,000 and 16,000 hours?

4-6 For the data in Problem 1-8, assume that a normal distribution model is a good fit. What is the 99 percent confidence interval for the mean and standard deviation?

4-7 For the data in Problem 1-8, use the Weibull distribution model and estimate the model parameters using a graphical method. Is the distribution close to the normal distribution? (Take $\Delta = 10$ kilocycles.)

4-8 For a specific item the following times to the first failure were recorded (in thousands of hours): 1.0, 1.1, 1.3, 1.41, 1.6, 1.65, 1.66, 1.9, 1.95, 1.97, 1.98, 2.0. 2.1, 2.2, 2.25, 2.28, 2.4, 2.42, 2.45, 2.6, 2.7, 2.75, 2.9, and 3.0.

    (*a*) Plot $R(t)$, $f(t)$, and $h(t)$ functions.

    (*b*) Check if the Weibull distribution is a good fit, and if it is, then find the parameters of distribution.

CHAPTER FIVE

STATISTICAL SIMULATION

A failure density function of a product is a function of failure density functions of components and interfaces and of the way these components are arranged into a system.

Let us suppose that both the reliability block diagram and the statistical properties of all components and interfaces corresponding to these blocks in this diagram are known. In general, the problem of finding the product's failure density function, and thus its reliability, does not have an explicit solution. Moreover, for a complex system and various statistical distributions for the components, even numerical solutions may be difficult to obtain with sufficient accuracy. An alternative is offered by the statistical simulation technique. If the input information into the simulation process is correct, the results of the simulation are equivalent to the results of the physical testing of the sample of products. The following example illustrates the concept of equivalency of physical testing and statistical simulation.

Example 5-1 Let us consider the reliability of a press-fit subjected to the torque variable in time. The system is simple, since it consists of one interface only.

Let us suppose now that for a specific press-fit the geometry is known (thus the material interference is known), the coefficient of friction is found experimentally (assume that it is done without destroying the press-fit), and the time dependence of the external torque is given. Also assume that all other geometrical and material properties are known. Then if the torque-carrying ability of the press-fit (strength) is denoted by S_T, which is a constant for a given press-fit, and the external torque (stress) is denoted by δ_T, which is a function of time, then the time to failure is found from the following equation (assuming that at some point the external torque becomes larger than the press-fit strength):

$$S_T = \delta_T(t) \tag{5-1}$$

Let us denote this time to failure as t_1. Another press-fit with different geometry and coefficient of friction has correspondingly different strength S_T, so that even for the same stress function $\delta_T(t)$, the time to failure would be different, $t_2 \neq t_1$. If such tests were carried out for a sample taken from the population, then a distribution of times to failure, $t_1, t_2, \ldots, t_n$, could be found. The solution of the problem of the press-fit reliability comes down to a need to measure the geometry and test a sample of the press-fit joints.

Let us consider now a situation in which the distribution of the shaft and hole diameters is known, as well as the distribution of the coefficients of friction. It is possible then to draw a pair from two populations (shafts and bushings) at random theoretically, and to find the corresponding material interference. Similarly, it would be possible to draw at random a coefficient of friction subject to a given distribution. As long as this information becomes available, the time to failure can be found from Eq. (5-1). This procedure can be repeated and another time to failure can be found as a result. In this way a distribution of the times to failure can be found.

The question is, what are the similarities and differences in these two approaches? The first approach is purely experimental: each of the two lots of shafts and bushings is drawn at random, measured, assembled, and tested. There is no prior knowledge on the properties of shafts and bushings or their interfaces.

In the second approach a shaft and bushing are also taken at random from two lots; however, the statistical properties of the lots are supposed to be known a priori, as are the properties of the interfaces. If our prior knowledge of the lots' (population) properties is adequate and the mathematical model of the contact pressure is correct, then there is no difference between the real life physical experiments and the numerical experiments because taking a draw from a real population or using a mathematical model of this population statistically produces the same result. The difference between the two approaches, however, comes when our prior knowledge is incomplete or nonexistent, or if the mathematical model is not adequate. In our case, in conducting the numerical experiment it was tacitly assumed that the coefficient of friction and the contact pressure, caused by the magnitude of metal interference, were independent. However, in reality, they are not independent. Thus if there is no prior knowledge on this dependency, the numerical experiment and the real life test would differ, according to the effect of the pressure-friction dependency.

Example 5-1 was given to make a point that a numerical experiment can substitute for real life tests, provided that the information concerning statistical properties of components and interfaces is known and the mathematical models of physical processes are correct. This is very important, since the more vague our knowledge about the system is, the less confidence we have on the result of a numerical experiment. The process of generating press-fits by choosing random shafts and bushings from the mathematical models of the corresponding distributions and subjecting them to the external loads is called statistical simulation. The process simulates the physical act of taking a shaft and a bushing from the two corresponding lots in a sense that a shaft and a bushing generated numerically belong to the corresponding physical populations.

In Example 5-1 the time to failure was determined from the deterministic equation Eq. (5-1). However, the torque as a function of time could be random as well. Then the time to failure should be found by generating a random function and finding a moment when equality Eq. (5-1) is satisfied. Again the statistical properties of the random load are supposed known.

Thus the simulation may be conducted if the statistical information about components and interfaces, external loads, and internal conditions is known. Let us suppose that this is the case. The next question is, how does simulation work? Let us assume for simplicity that the time to failure is a function of two random variables (in Eq. (5-1), it is a function of friction and metal interference):

$$t = g(x,y) \tag{5-2}$$

Consider, for example, the mean time to failure

$$\text{MTTF} = \int_0^\infty t f(t) dt \tag{5-3}$$

where $f(t)$ is the failure density function, which depends on the probability density functions (pdf) of two random variables, x and y, namely,

$$f(t) = \psi(x, y) \tag{5-4}$$

The integral in Eq. (5-3) can be written

$$\text{MTTF} = \int \int g(x, y)\psi(x,y) \, dx \, dy \tag{5-5}$$

where $g(x, y)$ is a known function and $\psi(x, y)$ is a joint distribution function of two random variables, x and y. Note that in the case of x and y being independent, $\psi(x, y) = \psi_1(x)\psi_2(y)$.

Let us assume that the integral is Eq. 5-5 cannot be analytically found. Then the following numerical process, the Monte-Carlo simulation, can be used. To make it more illustrative, let us consider Example 5-1 again.

Example 5-2 The press-fit strength is given by

$$S_T = Af\Delta \tag{5-6}$$

where f is the coefficient of friction, Δ is the metal interference, and A is a constant. The metal interference is, in turn, a linear function of two tolerances,

$$\Delta = \Delta_1 + \Delta_2 \tag{5-7}$$

Let us assume for simplicity that the torque is a deterministic function of time. The time to failure is then found from

$$Af(\Delta_1 + \Delta_2) - \delta_T(t) = 0 \tag{5-8}$$

For each value of f, Δ_1, and Δ_2, a value of t is found by solving Eq. (5-8) (see Fig. 5-1), which means that there is an inverted function of $\delta_T(t)$ such that

$$t = \delta_T^{-1}[Af(\Delta_1 + \Delta_2)] = g(f, \Delta_1, \Delta_2) \tag{5-9}$$

The MTTF is

$$\text{MTTF} = \iiint g(f, \Delta_1, \Delta_2)\psi(f,\Delta_1, \Delta_2) \, df \, d\Delta_1 \, d\Delta_2 \tag{5-10}$$

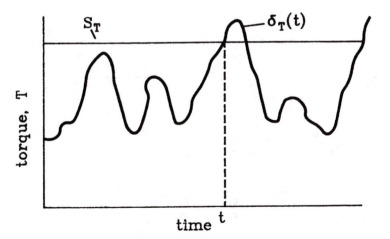

Figure 5-1 Illustration of time to failure.

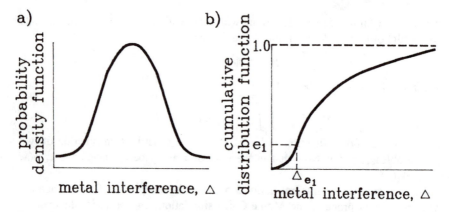

Figure 5-2 Illustration of simulation of a random variable Δ_{e1} for a given distribution function.

Since the function $g(f, \Delta_1, \Delta_2)$ often is not known analytically, the integral cannot be evaluated. In addition, the form of the distribution function $\psi(f, \Delta_1, \Delta_2)$ may complicate the situation.

Let us assume now that Δ_1 and Δ_2 are normally distributed. Then $\Delta = \Delta_1 + \Delta_2$ is also a normally distributed variable (see Fig. 5-2a). Let us also assume that the distribution function for f is given according to Fig. 5-3a, which may not have a specific mathematical model. For each of these two distributions it is possible to find the corresponding cumulative distribution function (see Figs. 5.2b and 5.3b). The ordinate of any cumulative distribution function changes between 0 and 1. The generation of a continuous random variable can be done in the following way. A random number having a uniform pdf is generated between 0 and 1, say, $0 < e_1 < 1$. This number is used as an ordinate of the cumulative distribution function, which leads to the corresponding abscissa (see Figs. 5-2b and 5-3b). Thus generated, two random variables, Δ_{e1} and f_{e1}, are substituted in Eq. (5-8), and corresponding time t_1 is found. Next, another random number is generated, $0 < e_2 < 1$, and the corresponding values $\Delta e_2, fe_2$, and t_2 are found. This procedure can be repeated n times; each repetition is called a run. Then an estimate of the MTTF in Eq. (5-10) is

$$\text{MTTF*} = \frac{1}{n}\sum_{i=1}^{n} t_i \tag{5-11}$$

Thus the estimate of the MTTF was found without an explicit representation of the function $g(f, \Delta_1, \Delta_2)$ and without performing the integration in Eq. (5-10).

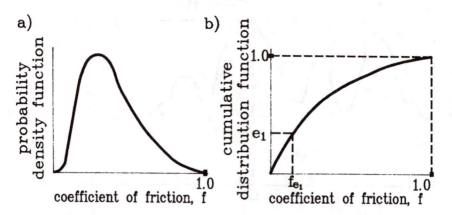

Figure 5-3 Illustration of simulation of a random variable f_{e1} for a given distribution function.

The next question is how MTTF and MTTF* are related. It is proven that if the number of runs tends to infinity, then

$$\lim_{n \to \infty} \text{MTTF}^* = \text{MTTF} \tag{5-12}$$

For a limited number of runs the standard error of the mean estimation is given by

$$\sigma_e^2 = \frac{1}{n} \sigma^2 \tag{5-13}$$

where σ is the standard deviation of the population (times to failure of press-fit joints in our case). From Eq. (5-13) it is seen that the error is decreasing as the square root of the number of runs.

The flowchart of the simulation program is shown in Fig. 5-4. The output of the program can be any statistic the design engineer is interested in: MTTF, variance, histogram, reliability, hazard functions, etc.

As can be seen from the above example, the Monte-Carlo simulation is based on experiments with random variables. Selecting random variables for a specified distribution is equivalent to selecting a sample from a population. The strength of the Monte-Carlo method is that it is not restricted by complicated analytical interrelationships between the variables describing a problem, and as a result, it allows all particulars of the phenomenon to be taken into account.

The Monte-Carlo method of simulation is analogous to experiments conducted on samples and as such provides an answer with the same degree of certainty. As with physical experiments, the larger the sample size, the more confidence one has in the result of simulation. In the case of the Monte-Carlo simulation, a larger sample translates into more simulation runs. The computer time of each run depends on the complexity of the system and can sometimes be formidable if many runs are required. Various variance-reduction techniques are developed to reduce the standard error, σ_e, while reducing the number of runs. One simple way to reduce the number of runs is to do as much analysis as possible theoretically and then apply the Monte-Carlo simulation to a reduced problem. Thus the complexity of the system is one factor affecting the efficient utilization of the statistical simulation.

Another factor is associated with the uncertainty of the mathematical model of the system, which follows from having incomplete or no statistical data about the failure rates of components and interfaces. Very often in statistical simulation the failure density functions for components are taken based on previous experience, mathematical reasoning, or expert judgment. The vagueness of the input information raises questions about the validity of the simulation results. The validity can be found only by checking the results of the simulation against real life data. The act of data comparison and analysis contributes to accumulation of both knowledge and experience in a specific field of engineering.

The Monte-Carlo simulation technique is applicable to problems where a scatter of parameters takes place. Two more examples, a small and a large size system, are given below.

Example 5-3 Consider the problem of degradation of a rubber seal due to temperature variations. The rate of degradation is a function of time over which the seal is exposed to a particular temperature. If the operating regime varies randomly, then so does the temperature of the oil at the seal. There are no analytical formulas relating the rate of degradation and the temperature. However, the experimental data relating these two factors might be available.

Let us assume that the degradation is measured in terms of the reduced resilience of the rubber seal and that resilience is measured in terms of the time needed for the seal to return to its original shape after a specific radial loading.

Let us take the simplest viscoelastic rheological element (Voigt body; see Fig. 5-5a), where the spring characterizes the elastic properties of the material and the dashpot characterizes the viscoelastic properties. Each property is described by the corresponding constants: E (modulus of elasticity) and λ (coefficient of viscosity). If the material is deformed to a level e_o and then suddenly released, the strain will decrease according to the relationship.

$$e = e_0 \exp(-Et/\lambda) \tag{5-14}$$

where the ratio $t_R = \lambda/E$ is called the restoration time.

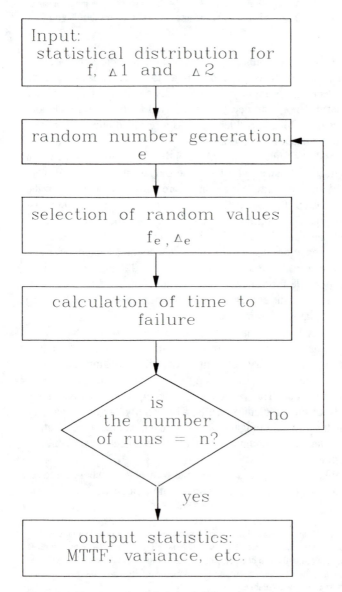

Figure 5-4 Flowchart of statistical simulation.

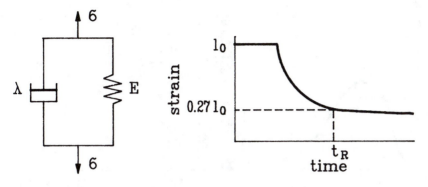

Figure 5-5 Viscoelastic element and corresponding length restoration curve.

When $t = t_R$, the initial strain drops by 63 percent (see Fig. 5-5b). Note that e becomes zero only at infinite time for this model. As viscosity of the material increases, the restoration time increases, i.e., it takes longer to restore the original state. Let us relate the degree of degradation with the coefficient λ characterizing the viscosity of the material. If at temperature T_i the material is able to withstand t_{max} hours of operation, then by considering this time as a maximum permissible restoration time, the maximum permissible coefficient of viscosity can be found:

$$\lambda^i_{max} = Et^i_{max} \qquad \text{at } T = T_i \tag{5-15}$$

An assumption should then be made about the damage model, i.e., what damage has been done to the material after $t < t^i_{max}$ hours of operation at temperature $T = T_i$. The simplest model is assumed here, namely, the degree of damage is proportional to the fraction of time of exposure relative to the maximum permissible time t^i_{max} at this temperature. Thus after $t < t^i_{max}$ hours of operation, the viscosity coefficient becomes

$$\lambda^i = \frac{t}{t^i_{max}} \lambda^i_{max} \tag{5-16}$$

Next, the linear cumulative model of damage is assumed, according to which the increase of the viscosity coefficient is represented as

$$\lambda = \sum \lambda^i \tag{5-17}$$

The Monte-Carlo simulation can then be used in the following way. It is assumed that the statistical data about the internal environment to which the seal is subjected is known, namely, the distribution of temperatures T and times of operation at a constant temperature t (see Figs. 5-6a and 5-7a, respectively). Alternatively, a joint distribution function $f(t,T)$ is known. For each distribution the corresponding cumulative distribution functions are constructed (see Figs. 5-6a and 5-7a). Then a random number is generated out of a uniform distribution between 0 and 1, say, e_1. For this random number the corresponding random variables T_{e1} and t_{e1} are found. From Eq. (5-15), λ^i_{max} corresponding to temperature T_{e1} is found, and from Eq. (5-16), λ^i corresponding to the time of exposure t_{e1} at temperature T_{e1} is found. Then another random number is generated, and the corresponding value λ^2 is found. This process continues until $\lambda = \sum \lambda^i = \lambda_{max}$. The time to failure is then

$$t_1 = \sum t_{ei} \tag{5-18}$$

The time t_1 is the outcome of the first run. After n runs, a distribution of times to failure is found, and various statistical measures can be calculated.

Example (5-3) allows some methodological generalization. The simulation methodology involves several distinct stages, although they may differ from one problem to another in terms of the following.

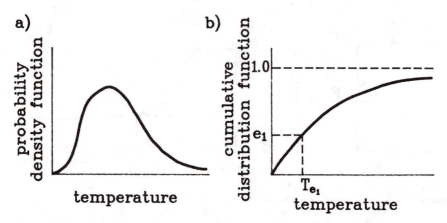

Figure 5-6 Illustration of simulation of a random variable T_{e_1} for a given distribution function.

1. Identification of a problem (degradation of a rubber seal due to the temperature variations).
2. Formulation of the problem (reliability of the seal for a given distribution of temperatures and times of exposure).
3. Formulation of the criteria of reliability (maximum permissible relaxation time t_R^p that, when exceeded, results in leakage).
4. Development of a mathematical model (linear viscoelastic model of the rubber).
5. Development of a damage model (linear additive model of damage).
6. Development of a statistical model (cumulative functions for the temperature and time distributions).
7. Development of a simulation code.
8. Simulation runs.
9. Analysis of results (hazard function, MTTF, etc.).
10. Recommendations (type of seal, type of oil, maintenance, etc.).

Some of these steps will be performed by a reliability engineer (writing a computer code, running and analyzing the results), others will be done by a design engineer (identification of a problem, recommendations), and still others will be done

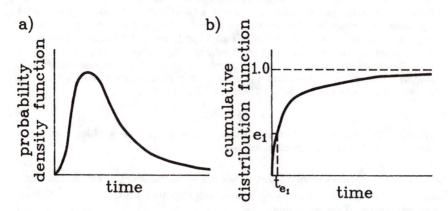

Figure 5-7 Illustration of simulation of a random variable t_{e_1} for a given distribution function.

together (formulation of the problem, criteria of reliability). The validity of simulation in Example 5-3 depends on (1) prior information about the distribution of times of operation at various regimes, (2) correctness of the mathematical model of seal behavior, and (3) correctness of the damage model. All these conditions make up the limitations for a possible extrapolation of the simulation results to other situations. However, within these limitations the simulation gives all the flexibility needed to make an optimal decision about the type of seal, type of oil, permissible operating temperatures, and time of maintenance.

Reliability analysis can be applied not only to predict the performance of a specific type of equipment, but also to predict the performance of systems comprising various types of equipment, including control systems and operators. The following example outlines the major steps in reliability evaluation of a gas-processing plant with emphasis on the role of a design engineer.

Example 5-4 Let us consider as an example a natural gas-processing plant.

Our concern here is with the reliability of the mechanical equipment of the plant, which includes, among other things, compressors, pumps, motors, heat exchangers, pressure vessels, pipes, and valves. The system is so complicated that even to identify items to be included in the reliability diagram, the combined efforts of a mechanical and a process engineer would be required, and to simulate the reliability of the plant, a reliability engineer would be needed.

Process plants frequently contain several identical parallel process trains, many redundant and standby subsystems and components. In some cases, parallel trains experience a variable demand. For example, the number of operating compressors can vary, depending on the amount of gas required. At the same time, since compressors fail randomly, the number of compressors available at any moment in time is random. To design such a system for maximum demand would mean to overdesign if maximum demand constitutes only a fraction of the overall time of operation.

Each equipment item or control instrument and even operator response (in terms of time) included in the reliability diagram must be analyzed in detail, namely, failure mechanisms, failure modes, MTTFs or MTBFs, and effects of a particular failure. For example, a water pump consists of two components: the pump itself and the motor. In turn, the pump consists of three physical components to be included in the reliability diagram: seals, bearings, and impeller rings. The motor consists of two physical components: the bearings and starter coil. Which mechanical items to include in the reliability diagram is decided by a mechanical and process engineer.

The statistical simulation of such a system may have several objectives:

1. To compute the average operating factor of the plant (the operating factor is defined as the average time that it operates at the design rate)
2. To compute the mean time between failures for critical components or subsystems
3. To compute the mean time between shutdowns
4. To study the costs of design alternatives.

A very important step in reliability simulation is gathering equipment failure data. For some items the statistics are known, for others it can be found in the literature, but in most cases (like pumps, compressors, valves) the data are simply not available. When no data are available, an expert estimation of failure density function is used. There are various approaches developed, but the simplest one is to use the triangular failure density function shown in Fig. 5-8. In this approach, experts are asked to identify an interval (a,b) in which they feel the failure will occur with probability close to 1. The most likely value c is also identified, and since the area of the triangle must be equal to 1, the coordinate of the maximum is $2/(b - a)$. The failure density is represented by two functions:

$$f(t) = \frac{2(t - a)}{(b - a)(c - a)} \quad a \leq t \leq c$$

$$f(t) = \frac{2(b - t)}{(b - a)(b - c)} \quad c \leq t \leq b \tag{5-19}$$

$$f(t) = 0 \quad \text{otherwise}$$

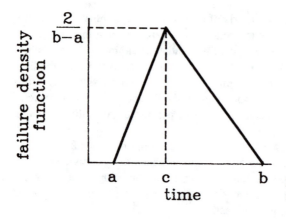

Figure 5-8 A triangular failure density function.

A reliability diagram plus data on failure density functions for each item in it, plus repair times, maintenance times, and downtimes for the whole system, constitutes a mathematical model of the system in question. In a simulation a computer is used to evaluate this model numerically over a specified period of time. The simulation starts by computing an initial failure date for each component in the system. Each failure time is a random sample drawn from the time-to-failure density function known for each component. To do that, the program generates a random number within the interval, say (a,b) in Fig. 5-8, which is used to compute a failure date. When failure dates for all components are computed, the first failure event is determined and stored. Since a component has failed, the program must sample the downtime density function for the failed component. The resulting date of repair is also stored. The program now examines the status of each component again (including the new one installed in place of the failed one) to find out the next item to fail, and a new date of failure is found and stored.

The program documents all failure times, repair times, etc., after each run. Each run generates a different set of random data and can be viewed as an experimental observation. The number of runs must be sufficient that average values of the operating factor, mean time between failures, and unscheduled shutdowns meet the prescribed level of confidence. If the mathematical model is valid, the increase in the number of runs results in more reliable statistics.

Computer simulation is a tool that can be used at various stages of product development and service. The simulation routines can be developed even at the conceptual stage. However, the depth of the system breakdown and the input information may vary during the design process, being upgraded when the information (analytical, experimental, published data) becomes available. In this way, a computer simulation program becomes part of the design process: using information from the design process and producing an output affecting the design process. In this design-simulation interaction, a design engineer plays the roles of customer and supplier at the same time. It is important to note that as a result of the involvement in the simulation analysis, the design engineer gets a much deeper understanding of the design in terms of its reliability.

PROBLEMS

5-1 The endurance limit for a shaft is found using information about the endurance limit for a specimen made out of the same material as the shaft and subjected to the same type of stresses. The relationship is usually given in the form

$$S_e^1 = K_s K_{sc} K_c K_t S_e$$

where K_s is the surface factor, K_{sc} is the scale factor, K_c is the stress concentration factor, and K_t is the temperature factor, all of which are random variables. Assuming that the pdf for all random variables are known, that is, $f(K_s)$, $f(K_{sc})$, $f(K_c)$, $f(K_t)$, and $f(S_e)$, outline the procedure of simulation for the endurance limit of the shaft.

5-2 In Problem 5-1, assume that the stress caused by the external load is also a random variable, the distribution of which is given by $f(\sigma)$. Outline the procedure of estimating the probability of shaft failure.

5-3 Assume that the reliability of a piston ring in an internal combustion engine is to be numerically simulated. Outline the various stages of the simulation methodology, and make an attempt to elaborate what is involved in each stage.

BIBLIOGRAPHY

The following are selected references for further reading.

Anderson, V. L. and R. A. McLean: *Design of Experiments,* Marcel Dekker, New York, 1974.

Ang, A. H. S., and W. H. Tang: *Probability Concepts in Engineering Planning and Design,* vol. 1, Wiley, New York, 1975.

Billinton, R. and R. N. Allan: *Reliability Evaluation of Engineering Systems,* Plenum Press, New York, 1983.

Bombas-Smith, J. H.: *Mechanical Survival: The Use of Reliability Data,* McGraw-Hill, New York, 1973.

Calabro, S. R.: *Reliability Principles and Practices,* McGraw-Hill, New York, 1962.

Dhillon, B. S.: *Mechanical Reliability: Theory, Models and Applications,* AIAA Education Series, Washington, D. C., 1988.

Dhillon, B. S. and C. Singh: *Engineering Reliability,* Wiley, New York, 1981.

Hahn, G. J., and S. S. Shapiro: *Statistical Models in Engineering,* Wiley, New York, 1967.

Hammersley, J. M. and D. C. Handscomb: *Monte-Carlo Methods,* Methuen, London, 1964.

Haugen, E. B.: *Probabilistic Mechanical Design,* Wiley, New York, 1980.

Kapur, K. C. and L. R. Lamberson: *Reliability in Engineering Design,* Wiley, New York, 1977.

Montgomery, D. C.: *Design and Analysis of Experiments,* Wiley, New York, 1976.

Morgan, B. J. T.: *Elements of Simulation,* Chapman and Hall, New York, 1984.

Nelson, W.: *Applied Life Data Analysis,* Wiley, New York, 1982.

Pieruschka, E.: *Principles of Reliability,* Prentice-Hall, Englewood Cliffs, N. J., 1963.

Sandler, G. H.: *System Reliability Engineering,* Prentice-Hall, Englewood Cliffs, N. J., 1963.

Shooman, M. L.: *Probabilistic Reliability: An Engineering Approach,* McGraw-Hill, New York, 1968.

Thoft-Christensen, P., and M. J. Baker: *Structural Reliability Theory and Its Applications,* Springer-Verlag, Berlin, 1982.

STATISTICAL TABLES

Table B-1 Standard Normal Distribution Function

$$P(Z \leq z) = \phi(z) = \int_{-\infty}^{z} \frac{1}{\sqrt{2\pi}} e^{-w^2/2} \, dw$$

$$\phi(-z) = 1 - \phi(z)$$

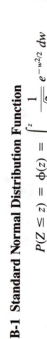

z	0.00	0.01	0.02	0.03	0.04	0.05	0.06	0.07	0.08	0.09
0.0	0.5000	0.5040	0.5080	0.5120	0.5160	0.5199	0.5239	0.5279	0.5319	0.5359
0.1	0.5398	0.5438	0.5478	0.5517	0.5557	0.5596	0.5636	0.5675	0.5714	0.5753
0.2	0.5793	0.5832	0.5871	0.5910	0.5948	0.5987	0.6026	0.6064	0.6103	0.6141
0.3	0.6179	0.6217	0.6255	0.6293	0.6331	0.6368	0.6406	0.6443	0.6480	0.6517
0.4	0.6554	0.6591	0.6628	0.6664	0.6700	0.6736	0.6772	0.6808	0.6844	0.6879
0.5	0.6915	0.6950	0.6985	0.7019	0.7054	0.7088	0.7123	0.7157	0.7190	0.7224
0.6	0.7257	0.7291	0.7324	0.7357	0.7389	0.7422	0.7454	0.7486	0.7517	0.7549
0.7	0.7580	0.7611	0.7642	0.7673	0.7703	0.7734	0.7764	0.7794	0.7823	0.7852
0.8	0.7881	0.7910	0.7939	0.7967	0.7995	0.8023	0.8051	0.8078	0.8106	0.8133
0.9	0.8159	0.8186	0.8212	0.8238	0.8264	0.8289	0.8315	0.8340	0.8365	0.8389
1.0	0.8413	0.8438	0.8461	0.8485	0.8508	0.8531	0.8554	0.8577	0.8599	0.8621
1.1	0.8643	0.8665	0.8686	0.8708	0.8729	0.8749	0.8770	0.8790	0.8810	0.8830
1.2	0.8849	0.8869	0.8888	0.8907	0.8925	0.8944	0.8962	0.8980	0.8997	0.9015
1.3	0.9032	0.9049	0.9066	0.9082	0.9099	0.9115	0.9131	0.9147	0.9162	0.9177
1.4	0.9192	0.9207	0.9222	0.9236	0.9251	0.9265	0.9279	0.9292	0.9306	0.9319

1.5	0.9332	0.9345	0.9357	0.9370	0.9382	0.9394	0.9406	0.9418	0.9429	0.9441
1.6	0.9452	0.9463	0.9474	0.9484	0.9495	0.9505	0.9515	0.9525	0.9535	0.9545
1.7	0.9554	0.9564	0.9573	0.9582	0.9591	0.9599	0.9608	0.9616	0.9625	0.9633
1.8	0.9641	0.9649	0.9656	0.9664	0.9671	0.9678	0.9686	0.9693	0.9699	0.9706
1.9	0.9713	0.9719	0.9726	0.9732	0.9738	0.9744	0.9750	0.9756	0.9761	0.9767
2.0	0.9772	0.9778	0.9783	0.9788	0.9793	0.9798	0.9803	0.9808	0.9812	0.9817
2.1	0.9821	0.9826	0.9830	0.9834	0.9838	0.9842	0.9846	0.9850	0.9854	0.9857
2.2	0.9861	0.9864	0.9868	0.9871	0.9875	0.9878	0.9881	0.9884	0.9887	0.9890
2.3	0.9893	0.9896	0.9898	0.9901	0.9904	0.9906	0.9909	0.9911	0.9913	0.9916
2.4	0.9918	0.9920	0.9922	0.9925	0.9927	0.9929	0.9931	0.9932	0.9934	0.9936
2.5	0.9938	0.9940	0.9941	0.9943	0.9945	0.9946	0.9948	0.9949	0.9951	0.9952
2.6	0.9953	0.9955	0.9956	0.9957	0.9959	0.9960	0.9961	0.9962	0.9963	0.9964
2.7	0.9965	0.9966	0.9967	0.9968	0.9969	0.9970	0.9971	0.9972	0.9973	0.9974
2.8	0.9974	0.9975	0.9976	0.9977	0.9977	0.9978	0.9979	0.9979	0.9980	0.9981
2.9	0.9981	0.9982	0.9982	0.9983	0.9984	0.9984	0.9985	0.9985	0.9986	0.9986
3.0	0.9987	0.9987	0.9987	0.9988	0.9988	0.9989	0.9989	0.9989	0.9990	0.9990

Selected Upper Percentage Points

Tail probability α	0.100	0.050	0.025	0.010	0.005
Upper percentage point $z(\alpha)$	1.282	1.645	1.960	2.326	2.576

Source: Reproduced in abridged form from Table 1 of E. S. Pearson and H. O. Hartley. *Biometrika Tables for Statisticians*, Vol. 1 (Cambridge: Cambridge University Press, 1954).

Table B-2 Upper Percentage Points of the Chi-Square Distribution

Values of $\chi^2(\alpha;r)$

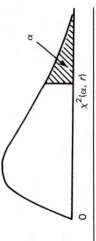

$\chi^2(\alpha, r)$

r	$\alpha = 0.995$	$\alpha = 0.99$	$\alpha = 0.975$	$\alpha = 0.95$	$\alpha = 0.05$	$\alpha = 0.025$	$\alpha = 0.01$	$\alpha = 0.005$	r
1	0.0^4393	0.0^3157	0.0^3982	0.00393	3.841	5.024	6.635	7.879	1
2	0.0100	0.0201	0.0506	0.103	5.991	7.378	9.210	10.597	2
3	0.0717	0.115	0.216	0.352	7.815	9.348	11.345	12.838	3
4	0.207	0.297	0.484	0.711	9.488	11.143	13.277	14.860	4
5	0.412	0.554	0.831	1.145	11.070	12.832	15.086	16.750	5
6	0.676	0.872	1.237	1.635	12.592	14.449	16.812	18.548	6
7	0.989	1.239	1.690	2.167	14.067	16.013	18.475	20.278	7
8	1.344	1.646	2.180	2.733	15.507	17.535	20.090	21.955	8
9	1.735	2.088	2.700	3.325	16.919	19.023	21.666	23.589	9
10	2.156	2.558	3.247	3.940	18.307	20.483	23.209	25.188	10
11	2.603	3.053	3.816	4.575	19.675	21.920	24.725	26.757	11

df									df
12	3.074	3.571	4.404	5.226	21.026	23.337	26.217	28.300	12
13	3.565	4.107	5.009	5.892	22.362	24.736	27.688	29.819	13
14	4.075	4.660	5.629	6.571	23.685	26.119	29.141	31.319	14
15	4.601	5.229	6.262	7.261	24.996	27.488	30.578	32.801	15
16	5.142	5.812	6.908	7.962	26.296	28.845	32.000	34.267	16
17	5.697	6.408	7.564	8.672	27.587	30.191	33.409	35.718	17
18	6.265	7.015	8.231	9.390	28.869	31.526	34.805	37.156	18
19	6.844	7.633	8.907	10.117	30.144	32.852	36.191	38.582	19
20	7.434	8.260	9.591	10.851	31.410	34.170	37.566	39.997	20
21	8.034	8.897	10.283	11.591	32.671	35.479	38.932	41.401	21
22	8.643	9.542	10.982	12.338	33.924	36.781	40.289	42.796	22
23	9.260	10.196	11.688	13.091	35.172	38.076	41.638	44.181	23
24	9.886	10.856	12.401	13.848	36.415	39.364	42.980	45.558	24
25	10.520	11.524	13.120	14.611	37.652	40.646	44.314	46.928	25
26	11.160	12.198	13.844	15.379	38.885	41.923	45.642	48.290	26
27	11.808	12.879	14.573	16.151	40.113	43.194	46.963	49.645	27
28	12.461	13.565	15.308	16.928	41.337	44.461	48.278	50.993	28
29	13.121	14.256	16.047	17.708	42.557	45.722	49.588	52.336	29
30	13.787	14.953	16.791	18.493	43.773	46.979	50.892	53.672	30

Source: Reproduced with permission from Table 8 of E. S. Pearson and H. O. Hartley, *Biometrika Tables for Statisticians.* Vol. 1 (Cambridge: Cambridge University Press, 1954).

Table B-3 Upper Percentage Points of the Student's *t*-Distribution

Values of *t(a;r)*

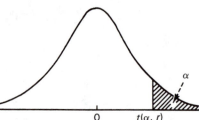

r	α = 0.10	α = 0.05	α = 0.025	α = 0.01	α = 0.005
1	3.078	6.314	12.706	31.821	63.657
2	1.886	2.920	4.303	6.965	9.925
3	1.638	2.353	3.182	4.541	5.841
4	1.533	2.132	2.776	3.747	4.604
5	1.476	2.015	2.571	3.365	4.032
6	1.440	1.943	2.447	3.143	3.707
7	1.415	1.895	2.365	2.998	3.499
8	1.397	1.860	2.306	2.896	3.355
9	1.383	1.833	2.262	2.821	3.250
10	1.372	1.812	2.228	2.764	3.169
11	1.363	1.796	2.201	2.718	3.106
12	1.356	1.782	2.179	2.681	3.055
13	1.350	1.771	2.160	2.650	3.012
14	1.345	1.761	2.145	2.624	2.977
15	1.341	1.753	2.131	2.602	2.947
16	1.337	1.746	2.120	2.583	2.921
17	1.333	1.740	2.110	2.567	2.898
18	1.330	1.734	2.101	2.552	2.878
19	1.328	1.729	2.093	2.539	2.861
20	1.325	1.725	2.086	2.528	2.845
21	1.323	1.721	2.080	2.518	2.831
22	1.321	1.717	2.074	2.508	2.819
23	1.319	1.714	2.069	2.500	2.807
24	1.318	1.711	2.064	2.492	2.797
25	1.316	1.708	2.060	2.485	2.787
26	1.315	1.706	2.056	2.479	2.779
27	1.314	1.703	2.052	2.473	2.771
28	1.313	1.701	2.048	2.467	2.763
29	1.311	1.699	2.045	2.462	2.756
30	1.310	1.697	2.042	2.457	2.750
40	1.303	1.684	2.021	2.423	2.704
60	1.296	1.671	2.000	2.390	2.660
120	1.289	1.658	1.980	2.358	2.617
∞	1.282	1.645	1.960	2.326	2.576

Source: Reproduced with permission from Table 12 of E. S. Pearson and H. O. Hartley, *Biometrika Tables for Statisticians,* Vol. 1 (Cambridge: Cambridge University Press, 1954).

WEIBULL DISTRIBUTION PROBABILITY PAPER

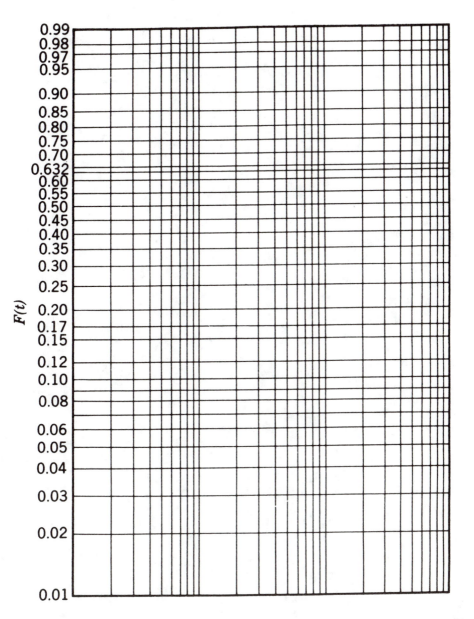

INDEX

Accelerated testing, 87
Alternatives in design, 4, 26, 31
Analysis of variance (ANOVA), 79

Conceptual design stage, 1, 26
Confidence level, 81
Confidence interval, 81
Confidence limits, 81, 103
Configuration, 1, 23
Coupling equation, 52
Cumulative function, 15

Design:
    alternatives, 4, 26, 31
    conceptual stage, 1, 26
    constraints, 1
    detailed stage, 2, 56
    preliminary stage, 43
    specifications, 2
Distribution:
    exponential, 99
    normal, 106
    Weibull, 110
Downtime, 5

Failure density function, 15
Failure rate, 15
Failures:
    early, 97, 98, 117
    random, 98, 117
    wear-out, 98, 117
Functional diagram, 27
Functional objectives, 1

Hazard function, 16
Hypothesis:
    alternative, 80
    null, 80

Level of significance, 81, 103
Life data analysis, 10, 114
Load-sharing, 36, 40

Maintainability, 5
Manufacturing stage, 4
Markov process, 36
Mean time between failures (MTBF), 20, 102
Mean time to failure (MTTF), 19, 100
Monte Carlo simulation, 121, 123

Percentile, 104
Population, 12
Preliminary design stage, 2, 43
Probabilistic design, 7

R-out-of-n configuration, 38
Redundancy, 34
   degrees of redundancy, 34
   levels of redundancy, 34
Relative frequency histogram, 14
Reliability:
   absolute level, 26, 45, 57
   component, 63
   definition, 11, 12
   relative level, 26, 45, 57
   system, 24
Reliability block diagram, 27
   components in parallel, 29
   components in series, 28
   components in series and parallel, 30
Reliability engineering, 5
Reliability tests, 74, 75

Safety factor, 45, 54
Sample, 11
Scatter, 6
Specifications, 2
Standby components, 41
Statistical dependence, 33

Statistical design of experiments, 76
Statistical hypothesis testing, 80
Statistical similarity, 32
Statistical simulation, 120
Stress-strength interference, 50

Tests:
   accelerated, 87
   acceptance, 74
   analysis of variance (ANOVA), 79
   Bayesian approach, 88
   blocking, 78
   development, 74
   qualification, 74
   randomized, 77
   reliability, 74, 75
   statistical design, 76
Time base, 32

Uncertainties:
   cognitive, 7, 11
   computational, 11
   downtime, 11
   external, 7, 10
   internal, 7, 10
   model, 11, 123
   physical, 7
   statistical, 11
Unreliability, 15
Useful life, 98

11-4-91

A000018670202